CAMBRIDGE COUNTY GEOGRAPHIES

General Editor: F. H. H. GUILLEMARD, M.A., M.D.

CARNARVONSHIRE

Cambridge County Geographies

CARNARVONSHIRE

by

J. E. LLOYD, M.A.

PROFESSOR OF HISTORY IN THE UNIVERSITY COLLEGE
OF NORTH WALES

With Maps, Diagrams and Illustrations

Cambridge :
at the University Press
1911

CAMBRIDGE UNIVERSITY PRESS
Cambridge, New York, Melbourne, Madrid, Cape Town,
Singapore, São Paulo, Delhi, Mexico City

Cambridge University Press
The Edinburgh Building, Cambridge CB2 8RU, UK

Published in the United States of America by Cambridge University Press, New York

www.cambridge.org
Information on this title: www.cambridge.org/9781107641624

First published 1911
First paperback edition 2013

A catalogue record for this publication is available from the British Library

ISBN 978-1-107-64162-4 Paperback

PREFACE

Though the author cannot claim a native's familiarity with the county of which he has here essayed to tell the story, long residence in it has given him an interest in its physical features and its history which he hopes is in some measure reflected in the following pages. Yet, pleasant as the task has been, it could never have been accomplished without the aid of many kind friends who have given him valuable expert guidance in unfamiliar fields. The chapter on Geology owes much to Mr Edward Greenly, F.G.S., and that on Natural History to Professor K. J. P. Orton. Help has also been given in various ways by Mr Harold Hughes, A.R.I.B.A., Mr G. J. Williams, one of H.M. Inspectors of Mines, Mr E. R. Davies, Secretary of the County Education Committee, Mr J. T. Roberts, Clerk to the County Council, Professor T. Winter, Dr J. Lloyd Williams and the Rev. T. Shankland.

J. E. LLOYD.

June 14th, 1911.

CONTENTS

ILLUSTRATIONS

ILLUSTRATIONS xi

MAPS

The illustrations on pp. 3, 9, 15, 22, 25, 27, 45, 57, 75, 77, 82, 98, 101, 109, 110, 112, 113, 119, 120, 135, 144, 148, 149, 152, 160 and 163 are from photographs supplied by Mr J. Wickens of Bangor; the rest from photographs supplied by Messrs F. Frith and Co., Ltd., of Reigate. The portrait of Sir John Wynne (p. 144) is reproduced from *The History of the Gwydir Family* by kind permission of Messrs Woodall, Minshall, Thomas & Co.

1. County and Shire. Meaning and Origin of the name Carnarvon-shire.

Shire is an Old English word which has long been used to denote a large division of the country, *shorn* off or separated by boundaries from the rest of the land. It is a *share*, to use another related word, of a wider area. From it is derived the Welsh word *sîr*, for which the equivalent *swydd* is sometimes found, denoting first, office, and then, the sphere within which an office is exercised.

Previous to the Norman Conquest of England, shire was the English term regularly used. The Normans, however, brought with them the name *county*, signifying the district ruled over by a count, and applied this to the shire, which they took to be a very similar institution. Hence it is that we speak indifferently of Carnarvonshire or the County of Carnarvon, of the Shire Hall and of the County Council. The two terms now have precisely the same meaning.

Carnarvonshire is the county which has its centre in the town of Carnarvon on the Menai Straits. The town has a much longer history than the shire, for there was a fort here, near Llanbeblig church, in Roman times, and it is this which the name commemorates. Carnarvon is properly "Y Gaer yn Arfon," the fort in Arvon, i.e. in the region fronting Anglesey.

All the shires of Wales are of much later origin than those of England. In the days of Welsh independence, the country was divided into cantrefs and commotes, each of much smaller extent than the English shire. The turning of Wales into "shire ground" was a part of the process of conquest. Carnarvonshire belongs to the earlier of the two sets of Welsh counties, namely, those constituted by Edward I after his defeat of Llywelyn, the last native Prince of Wales. It came into existence in March 1284, as the result of the Statute of Rhuddlan, sometimes known as the Statute of Wales. By this act three Welsh cantrefs and two commotes were grouped together under the Sheriff of Carnarvon and a new county was created.

2.   General Characteristics.   Position and Natural Conditions.

Mountains and the sea have made Carnarvonshire what it is, and have given it a character of its own that is hardly matched by any other county in Southern Britain. It has the longest coast-line of any Welsh county with the exception of Pembroke, and its mountains reach a height not attained elsewhere south of the Tweed. These are the natural features which made it in the Middle Ages the "strength of Venedotia" (cadernid Gwynedd), a sure retreat in time of invasion, and, with the adjacent isle of Anglesey, which it protected from attack, the chief seat of the power of the princes of North Wales. An expedition into Snowdon, as the English then styled the whole district, was a hazardous affair and, until Edward I used his fleet to second the efforts of his army, usually had little result.

View near Capel Curig

The county shares the mild and humid climate of the western coast of Great Britain and this, with the great extent of upland which it contains, has from the earliest times made it a region of pastoral rather than agricultural pursuits. The pastures of Eryri, or Snowdonia, were of wide renown in the time of Giraldus Cambrensis, who says that they were reputed sufficient to feed all the

flocks and herds of Wales. While the district has its
craggy heights and desolate moors, it has also rich valley
bottoms like Nant Ffrancon and the Vale of Nantlle,
affording the best of herbage. It still, notwithstanding
the changes effected by modern agriculture, preserves its
character as a stock-raising country, and such crops as
are grown are chiefly subsidiary to this end. The annual
yield of wheat is very small, but a fair quantity of barley
and turnips is raised for feeding purposes, while the crop
of oats is considerable. Little more than one seventh of
the surface of the county is under any kind of cultivation,
the rest being grass land, moor, rock, marsh, scrub, and
woodland.

Carnarvonshire is, in respect of situation, an essentially
maritime county, but it has never been remarkable for its
seaboard industries. Neither its seaports nor its fisheries
have ever been important elements in the development of
the district, though the rise of slate quarrying has of late
led to the opening of new ports and somewhat helped
the growth of the old. History shows, in fact, that the
Welshman has not, as a rule, taken kindly to the sea.
In spite of its persistent beating at his gates, he has always
regarded it as a strange and mysterious thing, best left to
itself, and he has never cared to reap from it the harvest
of food which it yields to the adventurous mariner. No
doubt the want of good natural harbours has been a
further handicap in Carnarvonshire, for much of the coast
is rocky and inhospitable, but this will not entirely explain
the comparative lack of sea-faring enterprise, which has its
origin in deep-seated racial habits.

The mountain, on the other hand, is the natural home of the Welshman and, while he has always roamed over it freely in the character of shepherd, he has of late come to still closer quarters with it as quarryman and miner. The mineral wealth of Carnarvonshire is very great, and during the nineteenth century the slate and granite quarries of the county have placed it in a new position as one of the leading industrial areas of Great Britain. Bethesda, Llanberis, and Nantlle annually produce great quantities of roofing slates; quarried, split, and dressed by a race of workmen whose skill is hereditary, and who have an honourable reputation throughout Wales for industry, thrift, intelligence, and love of learning. The quarry district extends across the county boundary to Festiniog in Merionethshire, with an outlier at Corris at the other end of that county. It is, industrially, the salient feature of North-west Wales and its influence is felt in every movement affecting the Welsh-speaking community, whether it be religious, literary, or educational.

Since the romantic movement at the end of the eighteenth century awoke men to an appreciation of the wilder beauties of nature, North Wales, and in particular Carnarvonshire, has been recognised as one of the most beautiful and attractive regions in the British Isles. It is not the mere height of its mountains, considerable as these are, which has drawn the sightseer to this region, but the rich variety of natural forms, the contrast of wood, rock, and water, picturesquely thrown together, as seen for instance in Nant Gwynant and at Bettws y Coed.

Among English counties, Bedfordshire and Hertfordshire
are nearest to it in area ; it exceeds Middlesex, Rutland,
and Huntingdonshire. Its greatest length is from the
headland of Braich y Pwll in Lleyn to Penrhyn Bay near
Llandrillo, a distance of about 55 miles ; lines running
north and south are much shorter, the longest being that
from the Great Orme's Head to Migneint mountain,
which is 25 miles, and that from Bangor to Portmadoc,
amounting to 20.

The county is of well-marked shape, projecting for
some 25 miles from the north-west corner of the Welsh
mainland in a south-westerly direction and ending with a
kind of snout at Braich y Pwll, off which lies the isle of
Bardsey, the westernmost point of North Wales. Its
long and tapering lines suggest the head of a greyhound,
with ears at Llandudno, eyes near Carnarvon, and nostrils
at Aberdaron. The river Conway makes a silver collar
and the Hiraethog region a sturdy neck.

Formed out of ancient Welsh tribal divisions, Carnarvonshire has, for the most part, natural boundaries clearly
marking it off from the neighbouring counties. North,
west, and south, it is hemmed in by the sea, its lengthy
coast-line stretching from the mouth of the Conway to
that of the Glaslyn. The channel known as the Menai
Straits, which for twelve miles winds between it and
Anglesey, is narrow, and in places less than a quarter of
a mile across, but the salt water sweeps it from end to
end at all states of the tide. On the eastern side the
main body of the county is separated from Denbighshire
by the river Conway, a noble river flowing through a

well-defined strath into a wide estuary.   It is only in the south-east corner, where the county abuts upon Merionethshire, that the boundary needs more careful definition, and even here it follows natural features.   Running up the Glaslyn to Cerrig y Rhwydwr it then keeps the course of the Dylif[1] to the cliffs of Yr Arddu.   From this point its line is that of the watershed, passing over the tops of

View on the Glaslyn

Cynicht, Moel Druman, Moel Farllwyd, Moel Penamnen, and Y Gamallt to the meeting place on the moorland of Migneint of the three shires of Carnarvon, Denbigh, and Merioneth.

[1] The district between the Dylif and Llyn Dinas, being the hamlet of Nanmor in the parish of Beddgelert, was transferred from the county of Merioneth to that of Carnarvon in 1888.

Three small portions of the county lie east of the Conway and, were it a mere question of simplicity of grouping, would be more suitably assigned to Denbighshire. All three were, in fact, included in the ancient cantref of Rhos, one of the main constituents of that county, but for various reasons have come to be associated with Carnarvonshire. The largest of the three was once the commote of Creuddyn, divided from the rest of Rhos by the marshy flat which extends from Llandudno Junction to Llandrillo. Popular tradition avers that in olden time it was along this level tract that the Conway found its way to the sea, and thus explains the connection of the peninsula with the county of Carnarvon. History has a much less startling tale to tell. Until the overthrow of Llywelyn by Edward I, Creuddyn was always reckoned a member of Rhos, and it was that monarch who, in constituting Carnarvonshire by the statute of Rhuddlan in 1284, modified the old arrangements by uniting the commote to the new county west of the Conway. His reason for doing so is not recorded, but it is to be remembered that at that time, and for long afterwards, Denbighshire had no existence, but was represented along the Conway by the marcher lordship of Denbigh, over which Edward had no direct authority. One may, therefore, presume that for military reasons he wished to keep full control of the mouth of the Conway and of the site of the ancient castle of Degannwy. If this was the case, Carnarvonshire owes to the military exigencies of the thirteenth century its power to draw to-day as a rating authority upon the valuable revenues of Llandudno.

A little north of Llanrwst, another portion of the county projects across the river into the county of Denbigh.   This is the township of Maenan in the parish of Eglwys Fach[1].   In 1284 Edward I, wishing to build a castle on the site of the Cistercian Abbey of Aberconwy, acquired this township from Earl Henry of Lincoln, the lord of Rhos, and gave it to the abbey in lieu of the old site and of lands in Creuddyn which he also took over. Maenan thus ceased to form part of the lordship of Denbigh and came under the authority of the Sheriff of Carnarvon.

The third detached portion of the county is the parish of Llysfaen[2], midway between Llandudno and Rhyl, which on the landward side is entirely hemmed in by Denbighshire.   This district appears also to have been granted away from Rhos at the time of the Conquest and to have been brought under the sway of the Carnarvonshire authorities.

4.   Surface and General Features.

No county in England and Wales has a more diversified surface than Carnarvonshire, where in the space of four miles one may travel from the meadows of Beddgelert, but little above the sea level, to the summit of Snowdon, 3560 feet high.   An ancient writer, describing

[1] A small portion, known as the Abbey, is extra-parochial.

[2] With this formerly went the township of Eirias, on the east side of Colwyn Bay, but in 1888 it was transferred to Denbighshire.

the part of the county which lies opposite to Anglesey, pictures it as made up of four parallel belts, rising tier above tier to the topmost crag of Eryri. First came the sea-board, then the champaign or plain country, next the wooded slopes, and last of all the rugged strength of the mountains. These divisions may conveniently be applied to Carnarvonshire as a whole. It has its 117 miles of sea-coast, sometimes bordered by majestic cliffs, but more often skirted by marshes and sand-flats. Next we have the region of farms, with tilled fields rising gently from the sandy beaches. If we fix the upper limit of this region at 500 feet, it will be found to account for about 250 of the 572 square miles which make up the land surface of the county. The woodlands clothe the lower slopes of the mountains from 500 to 1500 feet, or, to speak more accurately, did so in the ancient days, before sheep deprived the hillsides of their clothing of scrub and brushwood. This region occupies about 220 square miles and is now the seat of the slate-quarry industry, as well as a ranging ground for sheep and mountain ponies. Last of all, we come to the lonely mountain ridges, spreading themselves here and there and forming a long backbone for the county. Seventy square miles of Carnarvonshire lie above the 1500 feet level.

It will be well to consider the conformation of the county in greater detail. Its characteristic feature is the Snowdonian range, which stretches from Conway Mountain to Carn Boduan without a break save such as are afforded by mountain passes, and curves from sea to sea in the form of a great amphitheatre. The best view of this

Sychnant Pass

mountain chain is to be obtained from Anglesey, whence on a reasonably clear day the range may be seen at one glance from Penmaenmawr to the Rivals, forming a picture not easily effaced from the mind of him who has once looked upon it. There are several gaps, notably those formed by Nant Ffrancon, by the pass of Llanberis, by the Rhyd Ddu Valley, and by Bwlch Derwyn, but only the last of these is broad enough to affect in any way the impression of unity.

The higher ground begins at once at the mouth of the Conway with Conway Mountain and Penmaen Bach, which have their feet in the waters of the Irish Sea. After the slight gap of the Sychnant Pass (517 feet), a gradual ascent begins, which at 2000 feet reaches the bold and striking crest of Tal y Fan. A spur thrown out to the north-west touches the sea at Penmaenmawr, where granite rocks rise steeply from the beach to a height of 1550 feet. South of Tal y Fan the chain is slightly dented by the Pass of the Two Stones (Bwlch y Ddeufaen, 1403 feet), through which passed the old Roman road to Carnarvon; thereafter it pursues its upward trend, through Drum (2528 feet) and the rounded hump of Foel Fras (3091 feet), to the lofty summits of Carnedd Llywelyn (3484 feet) and Carnedd Dafydd (3426 feet). These are hardly inferior in height to Snowdon itself, but, despite some fine precipices, lack the advantages of the better known mountain in respect of isolation and grandeur of outline.

At this point we come to the first of the great Snowdonian passes, traversed by the old coaching road

The Precipices of Carnedd Dafydd

between Shrewsbury and Holyhead. This is formed by
the valleys of the Ogwen and the Llugwy, which run
north-west and south-east respectively from a watershed
1000 feet high. Singularly majestic scenery is the result
of the driving of this cleft through the heart of Snowdonia,
and lakes, torrents, and waterfalls enhance the appeal to
the imagination of the sheer walls and soaring pinnacles.
Another solid section of the great range follows, with
Tryfan (3010 feet), Glyder Fach (3262 feet), and
Glyder Fawr (3279 feet), summits no less notable
for their impressive beauty than for their actual height.
Glyder Fach[1], in particular, is a chaos of rocks, vividly
described by Charles Kingsley in *Two Years Ago*—
"a region where the upright lava ledges had been split
asunder into chasms, crushed together into caves, toppled
over each other, hurled up into spires, in such chaotic
confusion that progress seemed impossible."

The pass of Llanberis next intervenes and at Pen y
Gwryd (900 feet) branches off in two directions, eastward
to Capel Curig and the Llugwy valley, southward to
Beddgelert. The southern branch, now known as Nant
Gwynant (the true form is, however, Nanhwynen), serves
to open a connection with the pass of Cwellyn and Rhyd
Ddu and thus to isolate the great central mass of Snow-
donia, with the loftiest and most celebrated of its summits,
known to the Welsh as *Y Wyddfa*, the (Giant's) Tomb[2].
Snowdon is a cone of volcanic ash, with three deep

---

[1] Confused by Kingsley with Glyder Fawr.

[2] The Welsh distinguish *Eryri*, the region (Snowdonia), from *Y Wyddfa*,
the summit (Snowdon).

"cwms" carved out of its sides, and outlying peaks and ridges, such as Lliwedd, Yr Aran, Crib Goch (The Red Comb) and Crib y Ddysgl (The Dish Comb), which form the tremendous foreground of the famous prospect from the summit and give grace and dignity to the aspect of the mountain, as viewed from below.

The Pass of Llanberis

The third of the passes dividing the Snowdon range is that which runs from Beddgelert to Carnarvon, with Cwellyn Lake, stretched at the precipitous base of Mynydd Mawr, as its central feature. High mountain walls hem it on the western, no less than the eastern side, culminating in Moel Hebog (2566 feet), which looks down upon the village of the famous grave. Beyond this

Snowdon from Llyn Llydaw

point the ridge, running westward, is well defined ; it has the Vale of Nantlle in a fold on the northern slope, dips down to 450 feet at the gap of Bwlch Derwyn and reaches the sea in majestic cliffs at Garn y Môr, the westernmost of the three graceful peaks of the Rivals[1].

While the mountains of Snowdonia thus occupy most of the county, there is one portion of it which, as regards surface, has quite other features. The tongue of land which stretches out to the west and separates the bays of Carnarvon and Cardigan lies comparatively low, and the sea breezes blow across it unimpeded from shore to shore. Lleyn, to give it the name by which it has been known for centuries, has no heights of any importance, save for the isolated Carn Fadryn (1217 feet), and is a land of fertile fields and meadows, where the tillage of the soil is profitable, and the prospect is sylvan and alluring rather than bold and romantic. It has much in common with Anglesey, but has no Irish traffic to connect it with the outer world, and thus retains the old-world isolation natural to its position. Aberdaron, 15 miles from the nearest railway station, is the " Ultima Thule " of North Wales, a byword among all Welshmen for the remoteness of its situation and the rustic simplicity of its people.

---

[1] "The Rivals" is a corruption of the Welsh "Yr Eifl," meaning "The Forks."

5. Watersheds. Rivers. Lakes.

It is a natural result of the character of Carnarvonshire as a mountain mass almost surrounded by the sea that its rivers run short and rapid courses and are mountain torrents, for the most part, rather than rivers in the ordinary sense of the term. The Conway is the one true

The Conway near Trefriw

river of the county, with a broad valley, a winding, leisurely course, and a considerable estuary. If, however, the rivers are unimportant, the lakes are numerous and of great interest, and it is to them, quite as much as to its mountains, that the district owes its fame for natural beauty.

The chief watershed is that which parts the streams running north into the Irish Sea and Carnarvon Bay from those running south into Cardigan Bay.   From Braich y Pwll Head to the Rivals this line keeps close to the northern coast, save for one southward bend to Carn Fadryn. It then follows the main ridge of the Snowdonian range, passing through Llanaelhaearn, Bwlch Mawr, Graianog railway crossing, Garnedd Goch, Pitt's Head, and the Snowdon summit to Penygwryd.   Here it parts company with the central chain, making its way south to the neighbourhood of Cynicht, where it again bends to the east, forming for a while the county boundary. North of Penygwryd, the main ridge acts as the water-parting between the Conway and the streams which flow into Beaumaris Bay.

The Conway (Welsh "Conwy") is one of the best known rivers in the British Isles.   Its name is one which goes back to Roman times, and it must have fixed men's attention in every age as the moat drawn by nature for the defence of the Snowdonian fastnesses.   It was

> On a rock, whose haughty brow
> Frowns o'er old Conway's foaming flood

that the poet Gray placed the prophet-bard who foretold in sombre verse the fate of the successors of Edward I; while another famous poet, Edmund Spenser, speaks of

> Conway, which out of his streame doth send
> Plenty of pearles to decke his dames withall.

It rises in the moorland tarn of Llyn Conwy and, after winding its way past Yspyty Ifan, bends sharply to the

Fairy Glen, River Conway

left and sinks from the highland region of Mid Wales to
its own charming valley in the cascades of Bettws y Coed.
First, we have the Conway Falls, a famous salmon leap,
next the delicate beauty of the Fairy Glen, and lastly the
reaches at Bettws itself, where the level strath begins,
amid towering heights clad with trees to the very summit.

Bridge on the Dwyfor (Rhyd y Benllig)

Henceforward the Conway is a quiet, meandering stream,
which at Trefriw has its first taste of the sea, and thereafter is navigable for small craft, as it pursues its devious
way past the old Roman fortress of Caerhun and the ferry
(now a bridge) at Talycafn to the sands and marshes of its
estuary.

The other rivers of the county are of less note.   The

Ogwen, after passing through the lake of that name, spreads itself across the verdant flats of Nant Ffrancon, sweeps round the foot of the great Penrhyn quarry, and enters Beaumaris Bay through the glades of Penrhyn Park. The Saint[1] carries the waters of the Llanberis lakes to the Menai Straits at Carnarvon. The Gwyrfai, an ancient local boundary, flows from Llyn Cwellyn past

Aberglaslyn Pass

Bettws Garmon and Bont Newydd (a "new bridge" which has long ceased to justify its name) to a more westerly point on these Straits. The Llyfni is the stream of the Nantlle Valley, with a short course, but a substantial volume of water. In South Carnarvonshire, the Erch

[1] Seiont is not a true traditional form, but was invented by some antiquary, who thought it nearer to Segontium than the common Saint.

Llyn Idwal

helps to form the harbour of Pwllheli ; the Dwyfach and
Dwyfor meet, like two ancient lovers long parted from
each other, a little while before they finish their course
for ever ; and the Glaslyn, springing from the very heart
of Snowdon, passes through the lakes of Nant Gwynant
to Beddgelert and the far-famed pass of Aberglaslyn, where
in olden times, before the making of the Portmadoc
embankment, it found its goal in the expanse of Traeth
Mawr.

The county is remarkably rich in lakes and fully
deserves to be known as the Lake District of Wales.
They are found at all levels, from Llyn Dinas, near
Beddgelert, which is only 176 feet above the sea, to the
little tarn known as Llyn y Cwn (The Hounds' Lake),
under the shoulder of Glyder Fawr, which pours its
waters into the Devil's Kitchen at a height of 2500 feet
above sea level. The longest is Llyn Padarn, the lower
of the two Llanberis lakes, which has a length of two
miles, and other considerable pieces of water are Llyn
Ogwen, Llyn Cwellyn, Llyn Llydaw, Llyn Cawlyd and
Llyn Eigiau. Dulyn and Melynllyn, the " Black Lake "
and the " Yellow Lake," cradled in the naked rock,
supply Llandudno with water, while Cawlyd discharges
the same office for Conway and Colwyn Bay. While
some of the Carnarvonshire lakes thus play a most
necessary part in the life of to-day, others are bathed in
the light of romance and have become famous in story.
The dark cliffs which encircle Llyn Idwal are supposed
to have witnessed the murder of a prince of that name,
the son of the great chieftain, Owen Gwynedd, and popular

Dolbadarn

fancy (most probably without warrant) connects Llyn Geirionydd with the great sixth-century poet Taliesin, a modern monument on its banks keeping the tradition alive. Between the two lakes of Llanberis, on a rocky eminence, stands the ruined fortress of Dolbadarn, and another historic spot between two lakes is Bala Deulyn (The Junction of the Two Lakes) in the Vale of Nantlle, where Edward I once kept court, and where, long ages before, as legend runs, the bewitched Lleu, disguised as an eagle perched on an oak, fell into the lap of the enchanter Gwydion and by him was released from the spell which had kept him in miserable bondage.

6. Geology.

The different rocks which show themselves on the earth's surface are of various ages, and it is the business of the geologist to classify them and arrange them in chronological order. This he does by observing their bedding or stratification; the lowest in the series of beds is the oldest, the newest that which lies uppermost. Owing to the folding and consequent tilting of the beds, followed by the wearing away of the higher deposits, the older, low-lying strata may not only be exposed, but may even be lifted up into considerable mountain masses. Thus it happens that Wales, though one of the most mountainous parts of Southern Britain, contains some of the oldest geological formations, and Carnarvonshire shares very fully both characteristics.

	Names of Systems	Subdivisions	Characters of Rocks
TERTIARY	Recent Pleistocene	Metal Age Deposits Neolithic " Palaeolithic " Glacial "	Superficial Deposits
	Pliocene	Cromer Series Weybourne Crag Chillesford and Norwich Crags Red and Walton Crags Coralline Crag	Sands chiefly
	Miocene	Absent from Britain	
	Eocene	Fluviomarine Beds of Hampshire Bagshot Beds London Clay Oldhaven Beds, Woolwich and Reading Thanet Sands [Groups	Clays and Sands chiefly
SECONDARY	Cretaceous	Chalk Upper Greensand and Gault Lower Greensand Weald Clay Hastings Sands	Chalk at top Sandstones, Mud and Clays below
	Jurassic	Purbeck Beds Portland Beds Kimmeridge Clay Corallian Beds Oxford Clay and Kellaways Rock Cornbrash Forest Marble Great Oolite with Stonesfield Slate Inferior Oolite Lias—Upper, Middle, and Lower	Shales, Sandstones and Oolitic Limestones
	Triassic	Rhaetic Keuper Marls Keuper Sandstone Upper Bunter Sandstone Bunter Pebble Beds Lower Bunter Sandstone	Red Sandstones and Marls, Gypsum and Salt
PRIMARY	Permian	Magnesian Limestone and Sandstone Marl Slate Lower Permian Sandstone	Red Sandstones and Magnesian Limestone
	Carboniferous	Coal Measures Millstone Grit Mountain Limestone Basal Carboniferous Rocks	Sandstones, Shales and Coals at top Sandstones in middle Limestone and Shales below
	Devonian	Upper Mid } Devonian and Old Red Sand- Lower stone	Red Sandstones, Shales, Slates and Lime- stones
	Silurian	Ludlow Beds Wenlock Beds Llandovery Beds	Sandstones, Shales and Thin Limestones
	Ordovician	Caradoc Beds Llandeilo Beds Arenig Beds	Shales, Slates, Sandstones and Thin Limestones
	Cambrian	Tremadoc Slates Lingula Flags Menevian Beds Harlech Grits and Llanberis Slates	Slates and Sandstones
	Pre-Cambrian	No definite classification yet made	Sandstones, Slates and Volcanic Rocks

The oldest rocks in the county are those which are known as pre-Cambrian; it is believed that this series was laid down in an inconceivably remote age and then subjected to long ages of pressure and metamorphic change and denudation, until in course of time the Cambrian deposits, long considered the oldest sedimentary rocks, were spread over them. The coast from Braich y Pwll to Porth Dinllaen is made up of pre-Cambrian rock, and two other strips, running the one from Bethesda to Llanllyfni and the other from Bangor to Carnarvon, may possibly be of the same antiquity.

Next in order comes the Cambrian formation, occupying most of the region between Clynnog and Llanfairfechan; it is of great importance to the county as including, between masses of hard, dark grit, the valuable beds of slate which count for so much in the economy of the country. Within this area are the open quarries of Bethesda, Llanberis, Moel Tryfan, and the Vale of Nantlle. In addition to this main mass, the county also includes a corner of the great Merionethshire area of Cambrian rock, which crosses Traeth Mawr and shows itself around Tremadoc and Portmadoc. The Portmadoc rocks are rich in the fossils, or petrified animal and vegetable remains, which are characteristic of this epoch; the *Lingulella* or tongue-shell, for instance, is plentiful in the cliffs of Yr Ogof Ddu, near Criccieth.

Geologically, however, the bulk of Carnarvonshire belongs to the Ordovician system, which claims the main axis of the county from Aberdaron to Conway, and includes all the higher mountains except Elidyr Fawr.

The hard Ordovician grits and slates were deposited above the Cambrian rocks, and are well represented throughout Wales. But they do not exhaust the contents of this formation. Fortunately for the lover of striking scenery, enormous beds of lava and volcanic ash, as much as 3000 feet thick in some places, are interposed between the sedimentary strata, a process which provided the alternation of soft and hard rock which is necessary for the grandest effects of earth-sculpture. Here and there the sources of these lavas have been disinterred and remain as hard, granite-like bosses like Carn Boduan and the Rivals.

The next formation in order of time is the Silurian, widely diffused throughout Wales, but only represented in this county by a small section of the Denbighshire area, which crosses the Conway near the mouth of that river. After the close of the Silurian period, there came another very long interval, marked, so far as this corner of Britain is concerned, not by further deposits, but by vast changes in those already laid down. Great movements of the crust of the earth, whose general direction was from the south-east, crushed and folded the strata, induced in some of them the readiness to split evenly in certain directions which is the essence of a good slate, and probably raised them all above the level of the primeval sea so as to form a mountainous land.

By this time, though the beds of which the greater part of the surface of England is composed had still to be laid down, the material basis of Carnarvonshire was very largely completed. It is, indeed, likely that, after the process described above, most of the area was submerged beneath the sea of the Carboniferous period and received a deposit of limestone. But this was subsequently exposed to the air and worn away, except the two fragments which remain in the Great and Little Orme's Head and the coastal strip from Carnarvon to Nantporth, near Bangor. The Great Orme is a characteristic bit of limestone scenery, with its terraced cliffs, its many caves, and its short, smooth herbage—"a wide, wild stretch of splendid barrenness," as the late Bishop Walsham How called it, "a treeless expanse of grey rock-ledges and mossy turf, and low, weather-beaten gorse bushes." Since the Carboniferous period, the mass of rocks forming Carnarvonshire has been subjected to many influences: there has been more than one elevation and planing down of the surface, followed at last by the carving out of the existing hill and valley system. This last stage probably followed the time when the chalk deposits were being laid down over the greater part of England and came at the end of the Tertiary period, so that at the beginning of the Pleistocene Age, the Snowdonian region had attained something like its present form.

Yet, although the mountains of the district had taken shape and the now familiar glens and rivers had formed themselves, there were some important differences between Pleistocene and modern Carnarvonshire. The region was, in the first place, not maritime; the Irish Sea and St George's Channel were at this period dry land and the rivers of Snowdonia fed much greater rivers which wound their way through these now submerged tracts to a sea coast far out in the Atlantic Ocean. Anglesey was not

Great Orme's Head and St Tudno's Church

an island, but a tableland, with the Straits, perhaps, as a valley separating it from the highlands to the south. Welsh legend tells of a time when the Lavan Sands, between Beaumaris and Penmaenmawr, were a fruitful plain, overwhelmed by the resistless ocean in one great devastating sweep ; and in truth it is not so long ago, as geological time is reckoned, that this tract was submerged, only that the process was much more gradual than the popular tale would suggest.   Another feature of the Pleistocene Age in this part of the world was the existence, during a portion of the time, at any rate, of great glaciers like those of Switzerland and Norway, which were formed on the higher ground and slowly worked their way down the valleys.   According to the late Sir Andrew Ramsay, Snowdon was the centre of six glaciers, radiating in all directions, and the traces of their activity may still be observed in the surrounding region.   The surface of the rocks has often been rounded and smoothed by the action of the moving ice ; and scratches, all pointing, so far as the same valley is concerned, in the same direction, show where the embedded stones, like a graver's tool, made their mark as they were carried along.   Isolated blocks of stone, too, stranded in places far from their parent source, are evidences of the power of the glacier to transport to great distances the rocky masses which fell upon it from above.   Other proofs of glacial action are the rock-basins, such as those in which Llyn Dulyn and Llyn Llydaw lie, scooped out of the mountain side by the grinding ice ; and the glacial drift—the accumulation of sand, gravel, and clay, interspersed with great boulders, which covers many acres of the surface of the county. The drift was possibly deposited at a time when not merely glaciers held sway, divided by ridges of bare rock, but when the whole country was enveloped in a thick ice-cap, like that which veils in eternal snow all the peaks and ridges and clifts of Central Greenland.

Next came the waning of the glaciers. As they shrank into the recesses of the mountains, their last survivors left the concentric moraines which still form a belt across the mouths of Cwm Idwal, Cwm Glas, and similar glens. Since that time the forces of nature have been slowly but busily removing the traces left by the Glacial Period, cutting away the glacial deposits, silting up the lakes, splitting the ice-rounded rocks and bringing about their decay, so that the mountains are gradually assuming the aspect that is characteristic of ordinary land surfaces exposed to the action of the air.

7. Natural History.

Both the flora and the fauna of Carnarvonshire are of special interest, owing to the occurrence in this remote corner of the British Isles of many forms elsewhere rare or extinct. Careful records made by local observers over a long series of years have brought out many important facts in this direction.

The character of the flora of a district is determined not only by climate and altitude above the sea, but also by geological formation, since plants vary greatly in their ability to find nutriment in particular soils. The principle that each formation has its special flora is very well illustrated in Carnarvonshire, where there is a small but thoroughly characteristic limestone area, and an important Alpine region. The limestone region lies around Llandudno and includes the Great and the Little Orme's Head. Here are to be found the wild cabbage, the vernal squill, the rock Hutchinsia, the white beam tree, the wild madder, the rock samphire, the rock-rose, the catchfly, and the spotted cat's ear. The wild cotoneaster (*C. vulgaris*), of which this is the sole habitat in Great Britain, was discovered here in the eighteenth century, but it has almost disappeared of late years and threatens, like the royal fern, which has gone irretrievably, to fall a victim to the misguided zeal of collectors. The other interesting botanical area is supplied by the higher ground in the Snowdonian district. Owing to the calcareous volcanic ash which occurs so frequently at these high altitudes, there is a rich Alpine flora, characterised by the presence of many rare as well as beautiful plants. Such are the mountain spiderwort, *Lloydia*, found on the cliffs about Llyn Idwal and on the sides of Snowdon, but nowhere else in these islands, the purple and the snow saxifrage, the rose-root, and, among ferns, the green spleenwort and the very rare *Woodsia ilvensis*.

Both the limestone and the Alpine areas are noteworthy for the rarity of heather and foxglove. These, however, grow plentifully on the older sedimentary rocks of the county, where also brambles and ferns, and in particular the parsley fern (*Cryptogramme crispa*), are to be found in great abundance.

The mountains of Carnarvonshire still afford shelter and a breeding ground for the last remnants of several species of animals and birds which once graced the woodlands and commons of the British Isles. The relatively mild climate allows these species to live throughout the year in the most exposed situations.

Although the bear and the beaver have only left place-names to record their existence, and the wolf has been extinct for centuries, the wild cat[1] has only recently disappeared. The marten, polecat, and badger are still to be found, the two former however only in the most remote districts. The least sign of one or other species leads to a hot pursuit by keepers into the rocky recesses of the mountains where these creatures bring up their families in the summer. The fox and otter are relatively common, as are moles, shrews, all the rats and mice (voles), with the exception of the dormouse and harvest mouse. The black rat has only recently become extinct, and found one of its last retreats in this district. Besides the common hare, the blue or mountain hare, which becomes white in winter, is common on the mountains, where it has been introduced from Scotland. The coast is occasionally visited by seals.

Passing to the birds, the county is situated on both the well-known lines of migration, running north and south, and east and west from Great Britain to Ireland. In the winter time, many northern and continental birds come to take advantage of the mild climate. In the warm

---

[1] There is a Carreg y Gath (Cat Rock) near Pentir.

season the majority of British summer visitors reach Carnarvonshire, although frequently not in large numbers. A few of these which are common in England or further east, such as the nightingale, wryneck, lesser whitethroat, yellow wagtail, and certain warblers, are only rare stragglers to this county.

In the mountain passes one may yet watch the magnificent flight of the raven, the buzzard, and the peregrine, or hear the musical call of the still rarer chough (once plentiful at the mouth of the Conway)— the last isolated members of these species which have managed to escape the pursuit of the collector and the gamekeeper. The golden eagle still occasionally visits the crags which derive from it their name of Eryri. But although these and other birds are protected in Carnarvonshire, the law must be rigorously enforced for the next few years if these species are to be preserved in this county, or, in fact, in Wales.

In the winter the mountains and the high cwms are nearly deserted ; of the small birds the common wren seems alone able to brave the worst winters, and can be heard singing above 2000 feet, even in January. But in the spring great numbers of birds return ; meadow-pipits, ring-ouzels, dippers, grey-wagtails, stonechats, wheatears, sandpipers, curlews, snipes, missel thrushes, hedge-sparrows, cuckoos, and swifts are all common through the summer, whilst the golden plover, redshank, and others are not infrequently met with. Nowhere can the peculiar ways of the cuckoo be so readily observed as in this treeless upland region ; the female with her string of attendant males can easily be kept in sight on the bare hillside for a considerable time.

The contour of the country tends to keep many birds, mainly migrants, isolated in certain valleys or defiles, and thus one is able to observe with exceptional facility with what remarkable tenacity not only given species but given races of birds cling to certain places. For example, the garden warblers come in large numbers annually to the Aber valley; the pied flycatcher is common in the Conway and Llanberis valleys, but very rare in Nant Ffrancon: the whinchat is not seen in many suitable sites, but may be found year after year breeding in numbers in some small area in one of the valleys. The constancy of the return indicates that not caprice but some subtle attachment of the race to the actual place is the cause of the choice.

In the winter the shores are visited by vast numbers of waders and ducks. Most commonly the ringed plover, dunlin, redshank, curlew, oystercatcher, sanderling, and occasionally grey plover are to be seen. Of the duck, the wigeon is probably the most numerous, but mallard, shoveller, tufted duck, golden-eye, sheld-duck and pochard are continually met with. The goosander and redbreasted merganser visit the sea and inland lakes; the eider-duck and brent goose are not infrequently off the shore, and even the American hooded merganser has not only been seen but shot in the Menai Straits.

8.   A Peregrination of the Coast.

The coast-line of Carnarvonshire presents as charming a diversity as one could wish to see. For nearly a hundred and twenty miles it winds its way past beetling cliff and by sandy marsh and wooded height, scarce keeping its character for three miles together. If we commence our peregrination in the north-east, neglecting the detached portion at Llysfaen, we shall find that we enter the county near Llandrillo yn Rhos, where the sluggish Afon Ganol creeps through its sluices to the sea. Immediately afterwards we encounter the solid bulk of the Little Orme's Head, in Welsh Creigiau Rhiwledyn, a limestone mass which rises sheer out of the waters. It has caves which can only be reached by boat, and its steep cliffs have proved fatal to many an unheeding wanderer. Rounding the point, we are in the curve of Llandudno Bay, which stretches from the Little to the Great Orme in a bow of sand, once edged by green turf and haunted only by the seagull and curlew, but now beset by row upon row of tall lodging-houses and in the months of July and August dappled with a motley crowd of pleasure-seekers. Next comes the Great Orme itself, flat-topped, and from a distance appearing of no great interest, but wonderfully picturesque when one gets close under its shelving rocks and sees them rise in massive bulk tier upon tier above one's head. Since 1879 it has been possible to drive round the Head along a specially constructed road which gives easy access to every part, and its summit can also be reached by a funicular railway

Llandudno Bay

In Welsh it is known as Pen y Gogarth, from the ancient estate of the bishops of Bangor which lay on the south-western side and is now known as Gogarth "Abbey."

The Great Orme's Head is a peninsula and the isthmus which connects it with the mainland is a sandy flat, well known in Welsh literature under the name of "Morfa Rhianedd," i.e. The Ladies' Marsh. Tradition has nothing very explicit to say as to the origin of this name, but it connects the marsh with the history of Maelgwn Gwynedd, the sixth-century ruler of these parts attacked in so outspoken a fashion by the British writer, Gildas. Maelgwn died of the "yellow plague," and the legend relates that the plague came from Morfa Rhianedd in the form of a yellow monster, the mere sight of which, through the keyhole of the door of Eglwys Rhos, was the death of the proud tyrant. Maelgwn's castle of Degannwy was not far off, standing south of the marsh on a rocky boss set back a little from the shore of the Conway. It was a famous stronghold in later ages, in the fierce conflict between Welsh and English, but, in spite of its long history, the remains on the site are very scanty.

Near the busy railway centre of Llandudno Junction, the county boundary crosses the Conway. The river is here about half a mile broad, and one listens with due respect to the wonderful tale of Llywelyn of Nannau, who is said to have killed the parson of Llansantffraid by shooting an arrow at him across the water! The heights around Conway town are a picturesque setting for romantic stories of this kind; they rise in towering pinnacles from the sandhills of the Morfa, where in past ages many a stiff encounter took place between Welsh patriot and English invader, and in our own time many a mimic fight has been planned for the instruction of citizen soldiers quartered here for military training.   As we proceed westward, the coast grows more and more precipitous.   Penmaen Bach (The Little Headland) and

Penmaen Mawr

Penmaen Mawr (The Great Headland) both rise directly out of the sea, which beats incessantly upon their rugged flanks and often threatens to undermine the high road circling round them.   Between the two the mountain slopes are studded with the lodging-houses and cottages of the long straggling village of Penmaenmawr.   The Great Headland is one of the most striking landmarks of the district, forming the eastern buttress of the Snowdonian range as seen from the north. It is deeply scarred by granite quarries, which have the merit of providing a livelihood for the villages below, but also lie under the reproach of fast eating up the great prehistoric fortress beloved of antiquaries.

Beyond Penmaen Mawr, the coast takes yet another form. The foot-hills of Snowdon retreat a little distance from the shore and leave a belt of fertile meadow and plain, which intervenes between the mountains and the broad inlet of Beaumaris Bay. The Bay itself changes in character, its tumbling waves giving place at low water to the sandy expanse of Traeth Lafan (The Lavan Sands)[1], where in bygone days the coach route to Ireland ran for three miles across the sandbank to Beaumaris ferry. Fields line the edge of the salt water at Llanfairfechan and Aber, to be succeeded, when the mouth of the rushing Ogwen is passed, by the fair woods of Penrhyn Park. As Bangor is approached, the Bay narrows into the channel of the Menai Straits, which winds in and out between high wooded banks like a veritable river. It is indeed known to the Welsh as the river Menai (yr Afon Fenai) and, owing to its tidal peculiarities, behaves at times like a strong stream making for the sea. For the tidal wave from the Atlantic which makes high tide around the British coasts, entering the straits at Aber Menai, near Carnarvon, forthwith sweeps along the channel and is not checked at the other end for some two

---

[1] Probably so called from the species of sea-weed known as "laver" (Welsh "llafan").

Menai Straits : the Swellies

hours, that being the length of time the mass of tidal water has taken to surround the isle of Anglesey. As the tide ebbs, the set of the waters is in the opposite direction and the "river" runs from Bangor to Carnarvon.

Bangor pier and ferry mark the entrance of the Straits and thence it is but a little way to the Suspension Bridge, one of the earliest triumphs of British engineering skill, and so cunningly designed as, while amply fulfilling its purpose, to enhance rather than detract from the beauty of the scene in which it lies. The Tubular Bridge, carrying the London and North Western line to Holyhead, belongs to a somewhat later age; it has not the grace and lightness of its elder sister, but is not without a certain massive dignity. Between the bridges is the whirlpool known as the Swellies, born of the contest between the two opposing tides and decidedly dangerous to small craft. Its old Welsh name was Pwll Ceris, the Pool of Ceris, and its terrors were well known to the medieval bards, one of whom says that to attempt to cross it was no better than to put one's life to the hazard of a throw at dice[1]. Once out of the boiling waters, the navigator passed into the beautiful reach which curves towards Port Dinorwic and now divides the woodlands of Vaynol Park from the sunny lawns of Plas Newydd. At Port Dinorwic there is a slate wharf, where of old stood Y Felin Heli, the Saltwater Mill, of which the wheel was turned, not by a running stream, but by the ebbing Menai. From this point the straits grow broader and the banks slope more gently to the water.

[1] "Val rhoi hoedl ar y dis" (*Gr. Hiraethog*).

Carnarvon stands at the mouth of the river Saint, in a pretty situation, the natural charm of which has been heightened by the handiwork of man. At the back is the sturdy mass of Twthill, a fragment of pre-Cambrian rock which gives a touch of antique rudeness to the scene. In the foreground is the tidal estuary of the river, fringed on the one side by the hanging woods of Coed Alun[1] and on the other by the masts of the many vessels which here ship slate from the inland quarries. Between hillside and river lies the ancient town, with its lines of ancient walling, and, best of all, its magnificent castle, reckoned the finest ruin in the British Isles, and associated with many a romantic scene in the history of Wales. A few miles to the west the Menai meets the sea, in surroundings which are strikingly contrasted with those of its eastern mouth. Two long points of shingly sand form a pair of contracted jaws for the channel, while between them runs the narrow passage of Aber Menai, often crossed by the princes of North Wales in their journeyings from their fastnesses of Snowdon to their capital of Aberffraw.

From Aber Menai the Carnarvonshire coast bends southward in the shallow curve of Carnarvon Bay. The mountains are at first a distant background, separated from the sea by the well-tilled fields of Arfon, which drop gently down to the water level. Only the prehistoric hill-fort of Dinas Dinlle, half consumed by the waves, breaks the even sweep of the coast-line. But, as Clynnog is approached, the serried ranks of Snowdonia draw closer, leaving but a narrow margin for tillage and habitation,

[1] Alun, not Helen, is the old form.

and at last the triple-crowned Rivals forbid all further progress and the great cliffs of Garn y Môr, dropping a thousand feet to the sea, nobly flank the Snowdonian barrier. For the next few miles there is no lack of variety. Carreg y Llam (The Rock of the Leap), a noted haunt of seabirds, juts into the ocean; Porth Nevin and Porth Dinllaen are two delightful bays of clean, smooth

The Rivals (Yr Eifl)

sand hemmed in by lofty, grass-grown banks which curve gracefully around them. Porth Dinllaen is a natural harbour unsurpassed in North Wales and has often been discussed as a possible rival to Holyhead. But as yet no Irish trains run to this remote corner of Lleyn and, while the man of business may regret the failure to utilise natural advantages, the lover of nature can let the eye rest upon a scene of peace and harmony in which there is no discordant feature.

The coast of Lleyn is, for the most part, rocky and little visited. Its isolation makes it a favourite haunt of birds and beasts, such as the seal, the raven, and the shearwater, which are not to be found in the more frequented regions. When the end of the promontory is reached, the rocks become formidable cliffs, past which rushes with impetuous force the swift current of Bardsey Sound. This was the land terminus of the old pilgrim route to Bardsey, the "Isle of the Saints," where at St Mary's Chapel, now ruined, the wayfarers prayed for a safe passage to their destination before they committed themselves to the mercy of the deep. Close by is the Maen Melyn or Yellow Rock of Lleyn, another notable landmark of the district, once known throughout Wales for its brilliant colouring, as the poet Dafydd Nanmor shows when he uses it as a simile to describe the golden tresses of his Llio, the lady of the primrose and the honeysuckle locks[1]

Bardsey is the largest and best known of the islands of Carnarvonshire. In Welsh it is called Ynys Enlli, which has been explained as Benlli's Island, from an ancient mythical hero of Wales, better known in connection with the Vale of Clwyd. It lies two miles off the mainland, and is about $1\frac{1}{2}$ miles long and three-quarters of a mile broad. The northern portion is mainly occupied by a hill, which rises to a height of 545 feet, but slopes on the westward side to a little plain which has room for a

[1] Mae'n un lliw a'r maen yn Llŷn.

few farms. The southern limb, on which is the lighthouse, lies lower, while between the two is a neck of land on which is the ordinary landing place. It is, beyond a doubt, the most secluded spot in North Wales. Owing to the difficulty of the passage, communication with the mainland is uncertain and irregular, and many of the younger inhabitants have never left the island. In return, it enjoys some immunities; tithes, rates, and taxes are almost unknown, and it is not disturbed by the turmoil of parliamentary elections. It was this isolation which led to the settlement here in early ages of Celtic monks, or "saints," who desired to withdraw as far as possible from the strife and dust of the world and who are said to rest here in their thousands. Off the south-west corner of Bardsey there is a famous reef, known to medieval Welsh mariners as Ffrydiau Caswenan, where legend reported that Arthur's ship *Gwenan* had once been cast away.

The coast-line now trends to the east and brings one in a little while to the harbour of Aberdaron, strewn with bright-coloured pebbles, flanked by lofty cliffs, and pro-tected from the sea by the Seagulls' Islands (Ynysoedd y Gwylanod). Art has as yet contributed nothing (if we except the venerable Norman church) to the interest of this old-world village, but its natural attractions would make the fortune of a more accessible spot. The next remarkable feature of the coast is the Rhiw Mountain, rising steeply from the sea to a height of nearly 1000 feet; no sooner is this passed than we are at Porth Neigwl, stigmatised by sailors as Hell's Mouth, a long, straight stretch of sandy beach upon which the Atlantic rollers are driven by south-westerly gales with tremendous force.   At the eastern end of this the other promontory of South Carnarvonshire juts out into the sea, a level tableland, which shows a fine cliff-face at Y Pared Mawr (The Great Wall) and, a little further on, a waterfall

St Tudwal's Islands

tumbling into the sea (Pistyll Cim), but is otherwise unremarkable.

Rounding the Cim headland, we come upon the two islands of St Tudwal. Of these the smaller and western, the true name of which seems to be Ynys y Meirch (The Horses' Isle), is narrow and precipitous, and offers little foothold for anything but the lighthouse which now crowns it. The eastern islet has from time to time been inhabited; it contains the ruins of a chapel, utilised by a modern hermit, and it has some little extent of grazing ground for sheep. North of the islands is the excellent anchorage of St Tudwal's Roads, a shelter in rough weather for the small craft of Cardigan Bay, and especially welcome by reason of the absence of good road-

Llanbedrog Point

steads in the neighbourhood. The little river Soch makes its way into the Roads amid many sand-dunes, which give a very distinctive character to the little summer resort of Abersoch. Sand is greatly in evidence at this part of the coast, and it may be doubted whether there is a finer beach in Wales than that which, backed by tall sand-hills which shut out every trace of human habitation, stretches in an unbroken line from Abersoch to Llanbedrog Point.

At Llanbedrog the unfrequented portion of the county is left behind and the eye rests on familiar scenes. The coast curves round in a great arch of sand to Pwllheli, an ancient borough and market-town which of late has also

Moel y Gêst

become known as a watering-place. Two rivers, the Erch and Rhyd Hir, unite to form the harbour, and hard by, guarding the entrance, stands the granite boss of Carreg yr Imbill, or the Gimlet Rock. Winding streams, lofty fir trees silhouetted against the sky, and a glimpse of distant hills make up the fascinating background. There is little here to suggest the craggy strength of Snowdonia, whose summits lie far away on the horizon. But, as one approaches Criccieth, the features of the northern side of the county begin to reappear. The moors and hillocks of Western Eifionydd pass into the great spurs thrown southward by Moel Hebog; Criccieth Castle, dominating that pleasant little town, fronts the sea on its rocky pedestal, while between it and Portmadoc towers Moel y Gêst, a fit outpost of the battalions of Eryri. By the time that the mouth of the Glaslyn has been reached, we are in the heart of Snowdonia once more, and nowhere does the majesty of Y Wyddfa show to fuller advantage than from the embankment which carries us into Merionethshire.

9. Coastal Gains and Losses. Coast Protection.

The story of the coastal gains and losses of Carnarvonshire is one which it is not easy to tell simply, with due regard to established facts, for this is a favourite field of legend and romance, and old traditions have become so intertwined with modern arguments that it is well nigh impossible to disentangle them.

The popular account, going back into the middle ages, is that Beaumaris Bay was once a fertile plain, ruled over by Helig ap Glannog and watered by the river Ell, which discharged itself into the sea beyond Puffin Island. Helig's palace stood in the midst, to the north of Penmaenmawr, where its site is marked by the great stones, exposed at very low tides, which are still known as Llys Helig (Helig's Court).   One fateful day, there was a great inundation, which swept this fair land from end to end and for ever buried it in the bosom of the ocean.   Helig and many of his people were able to escape, and met to mourn their overwhelming loss at Trwyn yr Wylfa (Weeping Ness).   Another memorial of the catastrophe is alleged to be preserved in the name Traeth Lafan, which is for Traeth Aflawen, the Melancholy Shore[1].

Romantic as is this tale, it belongs to the region of folklore, and not to that of history.   It is known that Anglesey was already an island, divided from the mainland by a formidable tidal channel, when the Romans first appeared in this part of the country.   But in still earlier times, in the Neolithic Age, for instance, it is quite likely that this shallow bay may have been dry land. Sir John Wynne, in the seventeenth century, records the discovery here, at very low tides, of roots of oak and ash such as are found in the submerged forests of Cheshire. It is, in fact, probable that for ages the land has hereabouts been gradually sinking, and the process is still going on, to the great peril both of the highway and of the London and North Western railway line.   At one point below Penmaen Mawr, the railway company have abandoned the struggle with the ever-encroaching waves and, instead of seeking to barricade the line against them, have allowed them to pass freely under the permanent way, which here takes the form of a bridge.

[1] For a more probable derivation, see p. 44.

The road authorities also have found their task one of great difficulty where the road is not protected by the railway.

At the other end of the Menai Straits, too, it is clear that the sea has been gaining upon the land. Leland, writing in the sixteenth century, says that near Aber Menai "the sea hath eat up a little village on Carnarvon side," and the aspect of Dinas Dinlle would certainly suggest that it had for ages been exposed to the devouring fury of the waves. A little further south is Caer Arianrhod, a cluster of rocks about half a mile below low-water mark, which tradition has long pointed out as a submerged castle, once the brilliant court of the beautiful, but far from virtuous, Lady Arianrhod.

On the south coast, we have to set against these losses substantial gains won by engineering skill from the power of the sea. Until the beginning of the nineteenth century, the Traeth Mawr was what its neighbour, the Traeth Bychan, remains to this day, an expanse of marshy estuary, which stretched in one dead level, save for a few islands of rock, from Penmorfa to Llanfrothen and from Aberglaslyn to the sea. Sir John Wynne of Gwydir, a man of active and enterprising disposition, conceived the idea as early as 1625 of reclaiming this tract by means of an embankment, and endeavoured to draw into his design the famous Welsh engineer, Sir Hugh Myddleton of Denbigh. But Sir Hugh had other fish to fry and, moreover, appreciated better than his brother baronet the difficulties of the undertaking. Accordingly, the scheme slumbered until in the first decade of last century, an

Aberglaslyn Bridge

epoch of general development in North Wales, it was taken up by William Alexander Madocks, M.P. for Boston.   Having purchased the estate of Tanrallt, Mr Madocks in 1800 first cut off from the Traeth the Penmorfa section, occupying about two thousand acres, and turned into arable land what had before been a salt marsh.   On one corner of this reclaimed portion he built

Tremadoc

a new town, styled Tremadoc from its founder; he provided it with a central square, a market hall, an assembly hall, and a church, connected it by road with Beddgelert and Nevin, and transferred to it the fairs of the district. The issue showed how difficult it is to forecast the future of a city; in a few years Tremadoc was altogether eclipsed by its still younger sister, Portmadoc, which, without any

fostering care, grew into an important town as the result
of the ordinary working of economic causes.   In 1808
Mr Madocks obtained an act of parliament for the carrying out of his more formidable enterprise, the building of
an embankment across the mouth of the estuary and the
reclaiming of the land as far as Aberglaslyn.   This task

Portmadoc Harbour

he completed in 1811 at a cost of about £100,000[1]; the
"cob," as it is termed locally, links by a solid road the
two sides of the Traeth and protects about four thousand
acres of land from the ravages of the ocean.   In 1821 he
obtained a second act of parliament for the making of a
harbour at one end of the embankment and it was here,

[1] The poet Shelley, then living at Tanrallt, was one of the subscribers.

under the lee of Moel y Gêst, that the seaport sprang up which, as the result of the stimulus of the Festiniog slate trade, has reduced Tremadoc to insignificance.

Navigation around the Carnarvonshire coast is difficult for more than one reason. The region between the Great Orme's Head and the Menai Straits is one of extensive sandbanks, through which the navigable channels have a narrow and tortuous course. The pier constructed at Bangor in 1896 had to be carried out far into the Straits ere deep water was reached, and steamers plying from Beaumaris to Llandudno find it best to eschew the direct route and to pass close to Puffin Island.

Around the promontory of Lleyn the rock-bound coast is full of danger for mariners, although there are not many islets and sunken reefs. The sands of Portmadoc bar, too, need to be carefully negotiated. Against these dangers there is fully organised protection. The pilotage arrangements of the Menai Straits are in the hands of five Commissioners, whose authority extends from Penmon to Carnarvon Bar. The Royal National Lifeboat Institution has five stations in the county—at Llandudno, Porth Dinllaen, Abersoch, Pwllheli, and Criccieth, and its boats are frequently called upon for service. There are also three principal lighthouses, one under the control of the Mersey Docks and Harbour Board, and two under that of the Elder Brethren of Trinity House. That at Llandudno (managed from Liverpool) is situated on the north-west corner of the Great Orme's Head at a height of 325 feet above the sea, and has a light visible for a distance of 24 miles and helpful to the numberless vessels which sail from or make the mouth of the Mersey. Bardsey Island also has a lighthouse, the keepers of which share the solitude of the island folk and yet have no dealings with them, being usually ignorant of Welsh, the only tongue spoken by the natives. This lighthouse shows a fixed light, visible for 17 miles, and serves to indicate to

Llandudno Lighthouse

ships coming up St George's Channel the course they must take if they wish to keep clear of Cardigan Bay. The way to Pwllheli and Portmadoc is guarded by the St Tudwal's light, raised 151 feet above high water mark and flashing at fixed intervals, visible for a distance of 19 miles.

10. Climate.

Carnarvonshire belongs to the western climatic area of Great Britain, that in which insular, as opposed to continental, conditions have their fullest effect. It is greatly exposed to south-westerly winds, carrying much moisture, because they come from the sea, and this, with the mountainous character of the central portion, induces a high rainfall. On the other hand, it is protected from much of the cold, dry wind, blowing from the continent of Europe, which brings about a low winter temperature in the east of England and Scotland, and consequently it has an equable climate, not varying very greatly during the year. Its skies are apt to be cloudy, even when no rain falls, and less sunshine falls upon it than on the plains of the east.

The salient feature of British climate is the passage from the middle Atlantic towards Scandinavia of what are known as cyclones, i.e. air-whirlpools revolving round a centre, while at the same time travelling as a whole in a north-easterly direction. These are borne along as the result of a general drift of air in this direction, prevalent over the whole of Britain, except when an anti-cyclone— that is, an area of comparatively calm air, producing heat in summer and frost in winter—extends over this part of the world. The south-westerly breezes, coming from a semi-tropical region of the ocean, are moist and warm and, as they move north, bring with them rain, to which are added high winds during the passage of a cyclone.

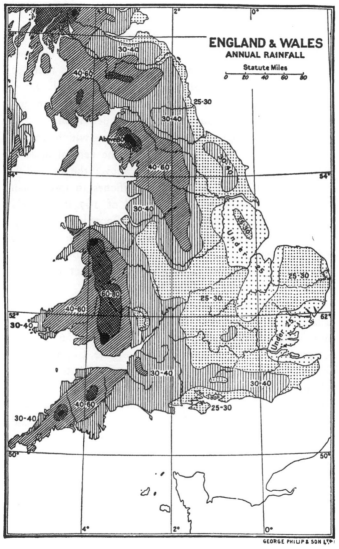

ENGLAND & WALES
ANNUAL RAINFALL

Statute Miles

(The figures give the approximate annual rainfall in inches.)

Striking against the mountain ranges of Wales, this warm current is driven upwards and in the process becomes cooled and parts with much of its moisture as rain; so that the Welsh highlands have a high rainfall for a double reason, their nearness to the source of rain and their capacity for producing further condensation.

Carnarvonshire well illustrates the causes which produce a wet climate or the reverse. Over the central mass of Snowdonia, for the reasons given above, the rainfall is very high, rising rapidly from 40 inches to more than 100 as we leave the coast. A fall of 167 inches was recorded on Crib Goch, a spur of Snowdon, in 1909. Skirting this central region is a belt of more moderate rainfall, including Criccieth, Carnarvon, and Bangor, with figures ranging from 30 to 40; in this tract the effect of the high background is only partially felt. Lastly, the outlying regions of the county, such as Lleyn and the Llandudno peninsula, escape altogether from this influence and have a rainfall of about 30 inches in the year, in common with a considerable tract of the English plain.

In compensation for the moisture of its climate, Carnarvonshire, like the rest of western Wales, enjoys mild seasons, the cold of winter and the heat of summer being alike assuaged by the neighbourhood of the ocean. The sea absorbs and gives out heat more slowly than the land : hence it is colder in summer and warmer in winter than the surface of the soil, and these qualities are communicated to the neighbouring lands, making an insular climate more equable than a continental one. Temperature is represented on maps by isothermal lines,

Crib Goch and Cwm Glas

drawn through places which have the same reading of the thermometer, and one of the most striking features of the British climate is the way in which in December and January these lines run north and south, instead of east and west, i.e. parallel to the equator, as we should expect them to lie, if nearness to the sun were the only element to be considered. The winter climate of North Wales is substantially the same as that of South Wales, Devon, and Cornwall, and it is shared by the Isle of Man and the west coast of Scotland. Within the limits of the county itself, the same rule holds good. In January, the coldest month of the year, the greatest warmth is found in Lleyn, where the average temperature is 42° Fahrenheit, as at Waterford, Holyhead, Cardigan, and Torquay. The next coldest region, with an average of 41°, is the Snowdonian, which goes with Dublin, the Isle of Man, Gower, and Exmoor. Colder still is the Conway Valley, which, with an average of 40°, is matched by the Pentland Firth, the Firth of Clyde, Central Wales generally, Swansea, and Cardiff. A similar tendency is to be observed in the summer months, though not to so marked an extent; in July, the Conway Valley and Snowdonia have the same average temperature, viz. 60°, as the Lake District and Carmarthenshire, while Lleyn, which is at this time the cooler end of the county, takes its place, with an average of 59°, side by side with Pembrokeshire, Anglesey, and the valley of the Tyne.

These figures, of course, take no account of height above the sea level. As soon as we begin to rise, temperature sinks, at the rate of 1° Fahrenheit for every

270 feet, and accordingly the slopes of the Snowdonian hills are much colder than the valleys which lie at their feet. Snow rests on the higher summits until May or June, particularly in sheltered gullies or where there is a steep northward face over which the sun has little hold, and no doubt it was the sight of these snowy cliffs gleaming in the distance which led English sailors, cruising in the surrounding seas, to give the whole group the name of Snowdon, or Hill of Snow.

11. People—Race, Language, Population.

All the great migrations into Britain have been, so far as is known, from the south and east, where access from the continent is easy. It was south-eastern Britain, with its low-lying plains, well fitted for agriculture, which always attracted invaders, and, as the result of their conquests, the other races were being perpetually driven north and west, into the highland and pastoral parts of the country. Accordingly, these regions now contain the oldest element in the population of the island, as is shown alike by history and by the physical characteristics of the modern inhabitants.

The great bulk of the dwellers in Carnarvonshire are Welsh-speaking Welsh. But it would be a great mistake to suppose that they are therefore of one race and one physical type, representing the Britons dislodged from the south-east by the English invasion. History shows that more than one race has been established from time to time among these hills, and observation leads to the belief that each has, in varying degree, transmitted its characteristics to the present population.

Students of prehistoric remains divide the ages which preceded the beginning of recorded history into the periods known as the Old Stone Age, the New Stone Age, the Bronze Age, and the Iron Age. During the first two, man had not learnt the art of casting metal into forms serviceable for domestic use and for war, and was therefore compelled to make his weapons and cutting tools of stone. In the Old Stone Age he could only chip and roughly fashion the stones of which he made use; the New Stone Age marks a great advance, for he had now acquired the art of grinding and polishing his material, so as to produce a fine cutting edge and a smooth surface. The next step forward was the discovery of bronze, a hard compound of two soft metals, copper and tin; this enabled him to improve his equipment greatly. Finally, at a period which in Britain did not long precede the dawn of history, and in Carnarvonshire may have been coeval with the Roman invasion, bronze was rejected as the metal of common use in favour of iron, and an epoch began which has lasted to our own day, for iron is still in request for every purpose requiring strength and solidity.

We need not here consider the civilisation of the Old Stone or Palaeolithic Age, for none of its remains have been found in Carnarvonshire, and it is most probable that the present-day population contains no descendants

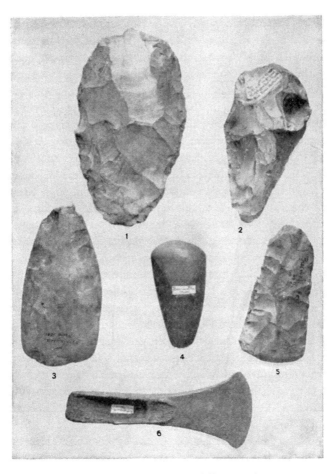

Weapons of the Stone and Bronze Ages

1 and 2 are Palaeolithic; 3, 4 and 5 Neolithic;
6 is a Bronze Age palstave

of the Palaeolithic race. But with the New Stone or Neolithic Age it is very different. The folk who then inhabited Britain have left abundant traces of their presence in the district; the cromlech was their characteristic burying-place; their weapons have been picked up from time to time; and in 1880 a cave in the Great Orme's Head was explored which contained skeletons of this people, with the perforated teeth of wild animals which they had worn for ornament. The Neolithic race was short in stature and of dark hair and complexion, being not unlike that which now inhabits Spain, and these are the features for which we most look, therefore, in endeavouring to trace their modern representatives.

The Bronze Age shows a totally different physical type; it saw the arrival in Britain of tall, fair-haired men, who not only had knowledge of the use of metals, but had different burial customs, interring the bodies of their dead (which were often burnt) in urns and stone coffins instead of in grave chambers. Bronze Age remains are as common in the county as those of Neolithic Age, and it is certain that the Bronze Age race invaded and settled in this part of the island. Indeed, it is believed that they were the people who introduced Celtic speech into Britain, the earlier or Neolithic folk having a language of a different type, which has become extinct.

Nevertheless, the facts collected by scientific observers as to the present physical type of Carnarvonshire would seem to show that the Bronze Age invasion did not, any more than the later Roman and English, displace the old population. For the prevailing type in this county, as also in Anglesey and Merionethshire, is short and dark. Admixture with fairer and taller races there has certainly been, for the short, dark type of Southern Europe is much more definitely so than that of any part of Wales. But comparison with England as a whole brings out the fact that there is more of the Neolithic element here than to the east, which is just what the history of Wales would lead us to expect. The Neolithic, or Iberian type, as it has been called from its Spanish air, is best represented in the secluded districts, such as Beddgelert, which have not been greatly influenced by foreign immigration.

According to the census of 1901, 12,165 of the inhabitants of Carnarvonshire spoke English only, 55,955 were familiar only with Welsh, while 49,346 were able to make use of both languages. Thus it would appear that, of the population over three years of age (to whom the language schedule was confined), about 90 per cent. were Welsh-speaking and only about 10 per cent. ignorant of that language. The county thus takes its place in that solid block of western Welsh counties, including also Anglesey, Merioneth, Cardigan, and Carmarthen, in which Welsh is spoken by about 90 per cent of the population. This predominance of Welsh shows itself in every part of the county, with the exception of the Llandudno area and—curiously enough—Bardsey Island[1], where the Welsh speakers form but 50 per cent. of the whole.

The total population of the county, as ascertained in 1911, was 125,049. It has increased continuously since 1801, when it was 41,521, except for slight decreases in

---

[1] For a partial explanation, see p. 61.

1881–1891 and 1901–1911. No doubt, the two main factors producing this increase have been the development of quarrying and the rise of seaside summer resorts. In respect of its population, Carnarvonshire holds the second place among the counties of North Wales, being a little outstripped by Denbighshire; of the South-Welsh counties Carmarthenshire and, of course, Glamorganshire have the advantage of it. It has about 220 persons to each square mile of its surface, but, in point of fact, this figure gives no clue to the real state of things. Extensive tracts of the county are as thinly populated as any region in Southern Britain, and the bulk of the population is concentrated along the coast-strip from Colwyn Bay to Clynnog, with the valleys of the Ogwen, Saint, Gwyrfai, and Llyfni, which are the region of the quarry villages. Moreover, the period of the year at which the census is taken, viz. the beginning of April, excludes altogether those who resort to the county for health's sake in the summer months, and thus a very important element in the economic life of the county is left out of account.

12. Agriculture.

It will be understood, from what has been said in previous sections, that various causes have united to reduce agriculture to a subordinate place in the economy of Carnarvonshire, and further, to make the tillage of the soil subordinate to the raising of stock. A moist climate has always been unfavourable to the growth of corn, and especially of wheat ; the rocky hill sides, with their thin coating of soil, do not repay cultivation, and of late other occupations have absorbed the energies of the worker. In 1901 the quarrymen and miners outnumbered the farmers and farm labourers of the county by about 5000.

Taking the figures of the year 1909, we find that 172,369 acres were under crops or laid out as permanent pasture, being less than half of the entire acreage of the county. The rest was mountain or heath land, most of it used for grazing, but not subjected to any kind of agricultural treatment. This equality of area between tilled and untilled land is not remarkable for a Welsh county ; there are several, indeed, which show a considerable preponderance of unreclaimed soil, such as Merionethshire and Brecon. But in England the only analogous case is that of Westmorland, and the contrast is striking with such a county as Lincolnshire, which has under crops and pasture 1,500,000 acres out of a total of 1,700,000.

On further analysis of the figures, it appears that permanent grass accounts for 122,811 acres and the various crops for only 49,558. Less than one-seventh of the surface of the county is, therefore, subjected to the plough, and a comparison may again be made with Lincolnshire, one of the great English corn-growing counties, where the arable land is nearly two-thirds of the whole area. The chief Carnarvonshire crop is oats, which occupies 10,606 acres ; barley rises to 4964 acres ; wheat is represented by the insignificant figure of 262. There are 3552 acres of potatoes and 2929 of turnips and swedes. No county in England and Wales, except Anglesey and Westmorland, produces less wheat.

According to the census of 1901, 7446 men and 857 women were engaged in agricultural work in the county, which amounted to about 15 per cent. of all having any active occupation. The number had not increased, but on the contrary had substantially decreased, since 1891.

The number of horses returned in 1909 was 9735, of cattle 53,956, of sheep 298,573, and of pigs 15,307. The most noteworthy feature of these figures is the high proportion of sheep, but this, while unusual from the English standpoint, forms a regular incident of Welsh rural economy. The small mountain sheep, which is the usual breed of the county, is remarkably hardy, and prospers with comparatively little care on the rough furze-clad moors and stony slopes of Snowdonia. Lambing in the early spring, and washing and shearing in the early summer, are the occasion of some anxiety to the breeder, but for the most part the sheep are left to themselves, and the fine flavour of Welsh mutton is largely due to the fact that they find their own food in the choice herbage of the mountain sides. Each farm adjoining the unenclosed mountain has its own "liberty," or sheep-walk, and also its special earmark, cut into the sheep's ear for purposes of identification; the farm sheep-dog, a marvel of animal intelligence, knows his master's sheep and can be trusted to collect them together, when this is necessary, with the minimum of supervision.

The horses include many of the mountain pony type, which run wild on the hills with the sheep and are

Black Cattle (University College herd)

known as merlins (Welsh "merlyn," *pl.* "merlod"). They are under twelve hands in height, and their small size makes them serviceable as pit ponies, for the drawing of loads in low-roofed colliery workings.

Carnarvonshire cattle are mostly of the black North Wales breed, which is a variety of the general Welsh stock specially associated with the island of Anglesey and the adjoining country. They have long horns, sleek and glossy coats, and graceful outlines; and besides being a good dairy breed, they produce excellent beef.

The county is to a very large extent one of small holdings. In 1909, out of 6356 agricultural tenements within its limits, 1508 did not exceed five acres in extent, being the largest number under this head in any county in Wales. There were only 19 farms which ran above 300 acres. Very many of the quarrymen have a little holding on which they grow potatoes and keep a cow, and the farms are often worked by the farmer and his relatives without the help of any hired labour.

The climate is not very favourable for the growth of fruit and green vegetables, and of Carnarvonshire it still remains true, as in days of Giraldus Cambrensis, that little use is made of the soil for orchards and gardens. The Welsh cottages, the slate roofs and solid stone walls of which seem to harmonise so well with the landscape in which they are set, have, as a rule, no bright profusion of country flowers to adorn them.

In the middle ages the district was well wooded, and the lower slopes of the mountains were covered with such a growth of copse and thicket as may still be found in the Aber valley. Drws y Coed, the "Pass into the Wood,"

at the head of the Nantlle valley, formerly led into a
thickly wooded glen, now bare of foliage and given up to
copper mining. Sheep have probably been the chief
agency in stripping the hillsides of their ancient glory,
for their nibbling of the young shoots would put a stop
to natural forest growth and, when once the mischief had
begun, the strong south-west winds would complete the

A Snowdonian Cottage

work, blighting the patches of woodland that remained
and preventing the growth of new timber. The Conway
valley, especially on its western or sheltered side, is the
only considerable wooded area in the county, though
there are also woods at Boduan, Glynllifon, Vaynol,
and Penrhyn. Some 11,716 acres in the county were
occupied by woods and plantations in 1905.

13. Industries—Mines and Quarries.

Quarrying is the chief industry of Carnarvonshire, as is evident from the fact that in 1901 by far the largest class of occupied males in the county consisted of persons employed in connection with mines and quarries, who numbered 12,098, while of these less than 500 were miners. This is the only industrial occupation which gives employment to a large number of hands, for there is now no manufacture or other important commercial enterprise in the county.

Mining is not very extensively carried on. The total number of persons returned in 1909 as engaged in or about the metalliferous mines of the county (slate mines being excluded) was 367, and the total value of the ore produced during that year was £17,483. Manganese is mined at Rhiw near Aberdaron, copper at Drws y Coed, iron at Bettws Garmon in Snowdonia, and lead and zinc at various points in the Conway Valley. In most cases the industry is at present at a low ebb, and many of the best-known mines of the county, such as the copper mines under the summit of Snowdon and around Beddgelert, have not for years been much worked. The author of *Beddgelert, its Facts, Fairies, and Folklore* (1899) quaintly says of the Snowdon mine : "it has been worked under several different names from time to time, but the name which shall give it fortune has yet to be given it."

The story of Carnarvonshire quarrying is very different. In 1909 there were 99 quarries of various kinds in the county and 10,169 persons employed in or about them, to which should be added six slate mines, occupying 331 persons[1]. Slate quarries took, of course, the first place, the total output of slate in the county for the year being of the value of £644,593. At the same time, limestone made a fairly respectable appearance with an output worth £42,611, and the various igneous rocks yielded £144,116. It may be said, therefore, that the mineral wealth of Carnarvonshire is turned to very good account.

The good qualities of North Wales slate for roofing purposes, its lightness, strength, and durability, have long been known, but it is little more than a hundred years since the development of the slate-quarrying industry into one of prime importance in this district. Richard Pennant, first Baron Penrhyn, was the principal mover in the matter. In 1782 he opened up the quarry at Cae Braich y Cafn in the Ogwen valley which has become the great Penrhyn Quarry, employing over 2000 workmen. In order to provide for the shipment of the slates, he connected the quarry with Bangor by means of a tramway, and formed at the mouth of the little river Cegin a capacious harbour, known as Port Penrhyn. The Penrhyn slates, which are of a deep blue colour, are known throughout the world for their merits as roofing slates, and are also in request for making writing-slates, monumental slabs, mantelpieces, and similar objects. The

[1] In these, situated near Festiniog, the slate is not worked in open quarries, but underground, in the Merionethshire fashion.

Dinorwic quarry, near Llanberis, was opened about the same time by Mr Assheton Smith; though not the best known, it is now the largest undertaking of the kind in the county, employing nearly 2800 men. The slates, which resemble those of the Penrhyn quarry in colour and general quality, are despatched by rail to the

Penrhyn Slate Quarries

quarry harbour at Port Dinorwic. Next to Bethesda and Llanberis, the most important slate centre is the Vale of Nantlle, where a number of quarries belonging to different owners employ in the aggregate about 1500 men. Nantlle slates are shipped from Carnarvon, with which town they are connected by the London and North

Western railway. There are also slate quarries on the slopes of Moel Tryfan, between Nantlle and Carnarvon.

Notwithstanding labour conflicts, some of which have been protracted, and a reduced demand for slate as roofing material, Carnarvonshire, with the adjoining region of Festiniog, still holds its position as a leading industrial area, and in respect of the production of slate is not approached by any other region in the British Isles. The slate-yielding rock is first detached from the mountain side in large blocks by blasting, then conveyed to slate-mills or cabins, where it is split and cut to the required shape by men who have remarkable manual dexterity in this work, and finally loaded into trucks which carry the slates to the port of shipment. Slate quarrymen are not paid according to time, but enter into "bargains" with the management, agreeing to turn a specified section of the quarry into slates for a stipulated sum. The workers on the face of the rock have "partners" in the splitting sheds, who share the bargain and deal with the slates sent to them by their comrades. Subsidiary industries which have grown out of the principal one are the making of school slates and the enamelling of slate for decorative purposes. So plentiful is the material that in many parts of the county a slate fence, bound with stout wire, is found to be the cheapest and most durable land-boundary.

Next in importance to the slate quarries come those in which the hard igneous rocks of the county, popularly known as granite, are worked. Their produce is chiefly used for paving, either in the form of small blocks or

Workers in the Penrhyn Quarries

"setts" or in loose chippings for macadamised roads. The two Penmaenmawr quarries treat in this way the diorite of that massive headland and employ between them nearly 1200 men. At Trevor, under the shadow of the Eifl mountain, some 350 quarrymen deal with a granite correctly known as augite porphyry. Another so-called granite is worked at the Gimlet Rock, near

Gimlet Rock

Pwllheli, the true name in this case being diabase, and here and there throughout the county small quarries may be found in which igneous rock is being picked out for road-making. In all not far short of 2000 men are occupied in this species of quarrying.

Lastly, one may mention the limestone quarries, employing about 500 hands. These are principally in the detached portion of the county lying east of Colwyn Bay. The two quarries at Llysfaen, where the Carboniferous limestone ridge juts out into the sea, are the chief source of occupation in that parish and supply an industrial background to the summer resort of Old Colwyn.

14. Shipping. Ports. Fisheries.

Although so largely sea-girt, Carnarvonshire has never had important maritime interests. During the period of rule of the independent princes, Aber Menai and Aberdaron were used as ports of embarkation for Ireland, but there was little general traffic along the coasts. The establishment of castles, with dependent boroughs, at Conway, Carnarvon, and Criccieth made a difference, setting up a demand for foreign commodities which led to a measure of intercourse by sea. But Beaumaris was for ages the chief seaport of this part of North Wales, being in fact the ancient customs centre of the whole coast from the Conway to the Mawddach. Conway, Carnarvon, Aberdaron, and Pwllheli were small trading centres doing a little trade with Ireland and Chester.

Shipping is not at present an important element in the economy of the county. Its maritime activity is in a large degree incidental to the land industries, that is to say, its ports are chiefly engaged in disposing of the products of its quarries. Bangor ships the slates of the Penrhyn quarries, Port Dinorwic those from Llanberis, Carnarvon those of the Nantlle district, and Portmadoc and Degannwy those of the Festiniog area. The slate wharves, except those of Carnarvon, are privately owned, but Pwllheli harbour and Bangor pier are the property of the respective corporations. The customs centre of the district is at Carnarvon, the area of control extending far beyond the limits of the county to Holyhead and Aberystwyth.

Pwllheli Harbour

Although the sea-board and the rivers of the county are well stocked with fish, it cannot be said that fishing is an industry extensively or profitably exercised by Carnarvonshire men. Only 143 persons were stated in 1901 to be employed in fishing in the county. The principal kinds of sea-fish are to be found in abundance around the coast; mackerel, cod, whiting, turbot, brill, plaice, sole, mullet and herring are all well represented; hake, haddock, and halibut alone being rare visitors in the district. In the Conway estuary, the sparling or smelt is found in great quantities in the spring. Most of these fish are caught by trawlers from Hoylake and the Isle of Man, though a little local fishing is carried on at Conway,

The Conway Estuary

Pwllheli, and Criccieth. The freshwater fish of the county are nearly all of the order of *Salmonidae*; salmon and salmon-trout are caught in the larger lakes and rivers, while trout are abundant in the smaller. The "torgoch" (i.e. red belly), or Welsh char, is found in Llyn Cwellyn and the Llanberis lakes.

15. History of the County to the Edwardian Conquest.

The Romans came into this district at a fairly early stage in their conquest of Britain, for they were bent on subduing Anglesey, where there was a large population, much under the influence of the Druids. Suetonius Paulinus was the first general to cross the Straits, but his victory yielded him no fruit, owing to the sudden rebellion of the British in his rear, and it was Julius Agricola who actually brought both island and mainland under the Roman yoke. The people on the shores of the Menai were a mixture of Goidels and Iberians, probably speaking an Irish dialect, and should be distinguished from the Ordovices of eastern North Wales. They became entirely subject to the Romans, but yet were not so wholly cowed as not to require careful watching. Two small forts were built in the district, the one at Caerhun, known as Conovium, the other at Carnarvon, known as Segontium, and it is certain that troops were kept in permanent occupation of these posts for very many years. A military road ran west from the great garrison town of Deva (Chester) and, crossing the Conway at Caerhun, pierced the Snowdonian range at Bwlch y Ddeufaen, whence it passed on to Carnarvon. Help was thus easily obtained in time of need from the headquarters of the Twentieth Legion on the Dee.

When, after an occupation of about three hundred and fifty years, the Romans withdrew from Britain, this corner of the island was held by men of Goidelic speech,

known to the Welsh as "Gwyddelod." There came
from the North about this time a chieftain named
Cunedda, whose followers were Welsh-speaking, and
whose victories made that tongue supreme in North
Wales. According to tradition, the Goidels of Snowdonia
fought obstinately ere they were overcome and made
their last stand in Nanhwynen ; this was probably under
Cunedda's grandson, Cadwallon of the Long Hand.
Cadwallon's son Maelgwn is known as one of the five
kings attacked by the sixth-century author, Gildas. He
was a powerful ruler, who bore sway over all North-west
Wales, and who seems to have made the rock of
Degannwy the centre of his realm. Gildas describes him
as one of the mightiest kings in Britain ; in his time,
therefore, the mountains of Carnarvonshire must have
given the law to no small part of Wales. He died about
550 and his name is still commemorated in that of Bryn
Maelgwn, near Degannwy.

By this time Christianity had reached the county, for
Maelgwn, though, according to Gildas, a man of many
sins, professed the Christian faith and had for a time
actually been a monk. Nothing is known of the first
preachers of the gospel in this neighbourhood ; probably
they came hither in the train of Cunedda and his family.
The earliest ecclesiastics whose names have been preserved are the "saints" (i.e. monks) of the sixth and
seventh centuries who founded the important and ancient
churches of the district. Thus Daniel, or Deiniol, set
up the monastery or "clas" of Bangor, Beuno that of
Clynnog, Trillo that of Llandrillo, and Hywyn that of

Aberdaron. These monasteries in time acquired a great deal of landed property, so that in the middle ages much of the land in this county belonged to the churches of Bangor and Clynnog. The pastures of the Great Orme belonged to the former and to the latter many a rich manor in Lleyn. Bangor also became the episcopal seat of Gwynedd, outstripping its rivals at Holyhead, Abergele, and Towyn.

The descendants of Cunedda and Maelgwn ruled this district for many generations. The main body of the county represents three ancient cantrefs, i.e. Arllechwedd, Arfon, and Lleyn, and one ancient commote, viz. Eifionydd[1]. Of these the first two remained in the hands of this family until the time of Edward I, there being but one break in the line of direct male inheritance, when Merfyn the Freckled succeeded in the ninth century in right of his mother. During a considerable part of the eleventh century, the old dynasty was out of possession; the strong and masterful Gruffydd ap Llywelyn, a doughty opponent of Edward the Confessor, was in power until his death in 1063, and then other alien rulers held Gwynedd for a time. But in 1081, as the result of a victory won in South Wales, Gruffydd ap Cynan, of the old stock, finally established himself upon the throne, which was afterwards held without interruption by his descendants. He had previously fought a battle in the county, at Bron yr Erw, near Clynnog, in 1075, which had turned out disastrously for him and had driven him back as an exile to Ireland, the land of his birth.

[1] For their situation, see p. 138.

After defeating his rival Trahaearn, Gruffydd had to contend with the Norman invaders of North Wales, who, in the reign of William II, nearly succeeded in conquering the whole district. Their first leader was Robert of Rhuddlan, who was, however, killed by the Welsh on the Great Orme's Head, not far from his castle of Degannwy, in 1088. Next came Earl Hugh of Chester, who built castles at Bangor and Carnarvon, as well as in the Isle of Anglesey. An unlucky expedition to the Menai Straits in 1098, in which his brother earl of Shrewsbury was slain, led Hugh to abandon his designs and to agree to accept homage from Gruffydd ap Cynan for Anglesey and the opposite coast. No doubt he realised that without a strong fleet, which the Normans did not possess, this quarter of Wales could not be subdued.

Gruffydd lived to a great age, shook off the yoke of the earl of Chester, held his own against the power of Henry I, and at his death in 1137 left to his sons a kingdom which stretched as far as the Dee valley. He was succeeded by his eldest son, known as Owain Gwynedd (1137-1170), who had a brilliant reign and whose power rested on the possession of Snowdonia. No foreign alarms now disturbed the glens of Eryri and, if the echoes of war were heard there, it was only as the result of civil strife. Owain and his brother Cadwaladr were often at variance, and the latter in 1144 brought a Danish fleet from Dublin to Aber Menai to vindicate his rights, but the quarrel was patched up and no conflict then took place. The two brothers were buried in Bangor cathedral, side by side with their father Gruffydd.

The halcyon days of Owain were followed by a period of strife among his sons and grandsons. Rhodri ab Owain gained the upper hand in the neighbourhood of Bangor and Carnarvon, and Giraldus Cambrensis found him in possession when he made his tour of Wales with the archbishop of Canterbury in 1188. But shortly afterwards there appears upon the stage the greatest of Welsh princes in the person of Llywelyn, son of Iorwerth ab Owain, who in 1194 won two notable victories hereabouts, at Aberconwy (Conway) and Porthaethwy (Menai Bridge), which launched him upon his triumphant career. From 1201 until his death Llywelyn the Great was undisputed master of Anglesey and Snowdonia, and from this centre carried on bold operations throughout North and South Wales. Once only did his power here receive a check, namely in 1211, when King John appeared with an army at the mouth of the Conway, sent his men to Bangor to burn the cathedral and carry off the bishop, and forced Llywelyn to make a complete submission to him at Aber. This was, however, but a temporary defeat; the quarrel between John and the barons enabled the prince of Gwynedd to recover himself, and henceforth his authority was unchallenged in these parts. He issued charters from Carnarvon, endowed an abbey of Cistercian monks at Conway, had a court (where his wife Joan died) at Aber, and castles at Criccieth and Degannwy. He spent his last hours in the precincts of Conway Abbey and was buried there in the monastic habit.

Llywelyn's son David had a short but stirring reign 1240–1246), which he closed at his father's court of Aber.

He had to bear the brunt of the hostility of Henry III, who in 1245 brought a great army to Degannwy. The campaign was as ill-starred as most of that monarch's warlike enterprises : provisions ran short, and on one occasion the English had the mortification of seeing their enemies loot a vessel laden with wine which, through bad steering, had grounded on the Conway Marsh, instead of under the castle walls, on the opposite bank. Henry gained nothing by this expedition, but David's death in the following year threw all into confusion, for he left no son, and Gwynedd west of the Conway fell to two of his young nephews, who were as yet no match for the power of the King.

The next stage in the history of Carnarvonshire was reached when in 1255 the nephews quarrelled, and at Bryn Derwyn, in the pass which leads from Llanllyfni to Dolbenmaen, the abler of the two, Llywelyn ap Gruffydd, won a signal triumph which was the beginning of a long and honourable career. Like his grandfather, the last Llywelyn found his chief support in the fastnesses of Snowdonia, of which he never lost his hold until the ill-fated year of his death. Degannwy belonged to a different region, to the ancient cantref of Rhos, and had not the same history ; Llywelyn did not win it from the English until 1263, when he reduced the garrison by famine, and in 1277, as the result of the first war with Edward I, it slipped from his hands again. But west of the Conway he was always supreme. Criccieth, Carnarvon, Dolwyddelan were among his courts ; Conway Abbey was benefited by his gifts. It was at Conway that the treaty of 1277 was signed, which left the prince with his original dominions, but took from him the four cantrefs between the Conway and the Dee. And it was on the Menai Straits, probably near Bangor, that the disaster befell the English in 1282, as they were crossing a newly-made bridge of boats, which almost turned the scale in Llywelyn's favour. Not until he fell, in a skirmish near Builth at the end of the year, did Carnarvonshire yield to the English power.

16. Later History of the County.

After the death of Llywelyn ap Gruffydd and the execution of his brother David, Edward I set himself to organise the government of the conquered country. In March, 1284, by the Statute of Rhuddlan, he created a number of new counties and of these Carnarvonshire was one. A sheriff and coroner were appointed for each one, and a Justice of Snowdon, with authority over the counties of Anglesey, Carnarvon, and Merioneth, while an exchequer was established at Carnarvon, which thus became the capital of North-west Wales. Another important measure was the building of castles; strong fortresses being erected at Carnarvon, Criccieth, and Conway. The building of the last named involved the dislodging of the abbey and its transference to Maenan, on the other side of the Conway. Around the castles boroughs were established—a new feature in the life of the district, for the Welsh had no towns in the days of their independence. Carnarvon, Criccieth, and Conway received their charters in 1284 and thus became little foreign colonies of traders, who supplied the needs of the men of the castle and, in time, did business also in market and fair with the Welshmen of the surrounding country.

Criccieth Castle

Although the intentions of Edward were good, he was often badly served by his ministers, and the oppression of these led to a general revolt in Wales at the end of 1294. A notable feature of this movement was the sudden attack upon Carnarvon, in which the sheriff of Anglesey was killed and the town and castle burnt to the ground. Edward was forced to undertake anew the conquest of Wales and he spent the Christmas of this year at Conway in far from cheerful circumstances. Fortunately for him, the Welsh were soon afterwards defeated in a pitched battle at Maes Madog (January, 1295), near Llanrwst, just outside the county boundary.

In 1301 Edward created his son and heir prince of Wales. The young prince had been born at Carnarvon in 1284, during the course of the conquest, but there is no foundation for the popular story that he was presented to the Welsh as their prince at the time of his birth. The revival at this time of the title borne by Llywelyn ap Gruffydd was very much of an afterthought, and was probably intended to give Wales a more independent position[1]. The prince came to Conway and there received the homage of the great men of Carnarvonshire. Among those who did homage at Flint was Sir Gruffydd Llwyd, lord of Dinorwig, who in 1322 led a rising against Edward II which had the sympathy of the men of Arfon, but was not successful.

Edward III was never prince of Wales, but in May, 1343, he raised his son, the famous Black Prince, to the dignity. On this occasion there was a formal investiture, the prince receiving the golden diadem, gold ring, and silver rod which were the ancient symbols of his office. The creation of a new prince (there had been none since 1307, when Edward II became king) led to many legal inquiries and one result was the compilation of the survey

[1] Llywelyn's predecessors were styled princes of *North* Wales, i.e. Gwynedd, and he was the first formally to assume the title of Prince of Wales. The crown recognised it in 1267.

of 1352, included in the MS. known as the *Record of Carnarvon*. In this there is a minute account of the state of the county in that year, similar to that contained in the Domesday Book of William I, and it appears from it that the Black Prince drew the same rents and services from the freemen and the serfs of the district as had been rendered to his Welsh predecessors. Pwllheli and Nevin owe their origin as boroughs to grants of the Black Prince.

During the fourteenth century the country became more peaceful and settled. The manufacture of a coarse frieze grew to be of some importance in the country and Carnarvon and Beaumaris exported the produce of fulling mills (Welsh, *pandy*), such as those at Castellmarch, Crewyrion, and Trefriw, although the trade never rose here to the height it assumed in South and Mid Wales. At the end of the century a period of strife is once more entered upon. The great rising of Owain Glyn Dŵr, which for ten years taxed all the resources of the English crown, affected Carnarvonshire, in common with every other district in Wales, and some of its notable incidents took place within the county. Such were the surprising capture of Conway castle by the Welsh on Good Friday, 1401, when the garrison were at church in the town, the attack upon Carnarvon by a Breton force in 1403, and the signature at Aberdaron in February, 1406, of the famous agreement between Glyn Dŵr, Northumberland, and Mortimer to divide England and Wales between them[1]. The shores of Merioneth and Cardigan were,

[1] Shakespeare, following Holinshed, wrongly locates this incident at Bangor.

however, the chief centre of Glyn Dŵr's power and, except for a brief tenure of Conway in 1401, the Carnarvonshire castles were never in his hands.

The remainder of the story of the county must be told briefly. Despite the overthrow of Glyn Dŵr, neither the Welsh language nor the Welsh spirit died out in this district and many Carnarvonshire men, under Sir William Stanley, had the satisfaction of helping at Bosworth to win a crown for the Welshman, Henry Tudor. The country was far from well governed: "so bloody and ireful were quarrels in those days," says Sir John Wynne of Gwydir in narrating the history of his house, "and the revenge of the sword at such liberty, as almost nothing was punished by law, whatsoever happened." Sir John's ancestor, Maredudd ab Ifan, left his old home at Cesail Gyfarch in Eifionydd and removed to Dolwyddelan, for, said he, "if I live in mine house in Eifionydd, I must either kill mine own kinsmen or be killed by them." The Act of Union and other measures of Henry VIII were intended to remedy these disorders by bringing about a closer connection with England. The Welsh counties were for the first time included in the system of English parliamentary representation, and in 1541 two men bearing famous names appeared as first representatives of the county of Carnarvon at Westminster, Sir Richard Bulkeley being the knight of the shire and John Puleston M.P. for the boroughs.

In the great Civil War, Carnarvonshire was predominantly royalist, like most districts in Wales, and the castles of Carnarvon and Conway, having been fortified for the king, held out until the collapse of his fortunes. Carnarvon was taken in June, 1646, by General Mytton, who had the help of Colonel Glynne of Glynllifon and Sir William Williams of Vaynol. Conway had been greatly strengthened by John Williams, Archbishop of York, of the Penrhyn family, but the place was afterwards transferred to Sir John Owen of Clenenneu and it was he who

Carnarvon Castle

surrendered it in November, 1646, to the parliamentary forces. The first and most important stage of the Civil War was now over, but the county also played a part in the struggle of 1648, when the royalist party made an attempt to reopen the war. Sir John Owen raised a force for the imprisoned Charles, which was, however, met and defeated on June 5th near Llandegai, Sir John and many other prisoners falling into the hands of the victors.

Through such vicissitudes did the district pass ere it attained the peace and prosperity of recent times. During the eighteenth century it was a rustic, secluded area, in which the tides of life flowed but sluggishly; the nineteenth saw a great development of industrial energy, of sightseeing traffic, of religious and literary and educational effort.

17. Antiquities.

In an area where there have been very few racial changes and revolutions, and where reverence for the past has always been strong, one might expect to find abundant memorials of bygone ages, and such is in fact the case. Carnarvonshire is rich in prehistoric antiquities, which until lately have been preserved by the popular feeling that it was unlucky and dangerous to interfere with them.

Of these relics of the past none are more striking than the cromlechs of the county. These wonderful erections were undoubtedly sepulchral, and constructed by the Neolithic folk for the burial of their more illustrious dead. The leading feature of the structure is the roof, always formed by a great stone slab or capstone, which rests upon three or more stones placed vertically and known as supporters. Cromlechs are known by many names, such as "Arthur's Quoit" (a favourite description), "The Altar" and "The Shelter Stone" (the original meaning of *cromlech*), but these names are the product of

7—2

popular fancy and tell us nothing of the real purpose of the structures.

The county contains about a dozen cromlechs, in

Rhos Lan Cromlech, near Criccieth

various stages of ruin.   Some have but the capstone left, the supporters having gone within recent memory.   The best preserved are that on the Great Orme's Head, the two near Clynnog, the two near Criccieth, and that at

Cefn Amwlch in Lleyn.   In the case of the two at Four
Crosses and Porthlwyd, the capstone has slipped off the
supporters.   No remarkable discoveries have been made

Cromlech at Cefn Amwlch

in connection with any cromlech in this county, but that
at Fach Wen, near Clynnog, is noteworthy as having
small cup-like hollows on the upper surface of the cap-

stone, a feature known elsewhere, but very unusual in Wales.

Another type of megalithic monument is furnished by the standing stone or *maen-hir*, often elaborated into a stone avenue or circle. These remains were at one time to be found in great abundance on the moors above Llanfairfechan and Penmaenmawr, but many have been

Stone Circle, Penmaenmawr

broken up or have ceased to be recognisable. Of the two standing stones which gave its name to Bwlch y Ddeufaen, one is now prostrate, and stone circles hereabouts described by the older antiquaries have become difficult to find. But the circle known as "Y Meini Hirion" (The Long Stones), at the back of Penmaenmawr, still makes an impressive picture; about ten of the stones remaining in position on the well-defined circumference of a circle measuring about 80 feet across. The date and purpose of these relics of olden time have not yet been determined, but the present tendency is to regard them as late Neolithic. They were certainly in position long before Caesar's time.

In the Bronze Age bodies were cremated and the ashes buried in earthenware vessels beneath barrows or cairns. Burials of this kind have come to light at Penmaenmawr and at Llystyn, near Dolbenmaen; and the many "carneddau." (cairns) scattered over the mountainous part of the county were, no doubt, Bronze Age burying-places.

Prehistoric forts or camps of refuge are well represented in the county. That on the summit of Tre'r Ceiri, the eastern peak of the Rivals, is one of the finest in Great Britain. It stands 1600 feet above the sea and commands a wide view of Arfon and Eifionydd. The wall is a solid rampart of unmortared stones, carefully fitted together, with gates and a platform for the defenders running along its inner side. Within the enclosure are the remains of a great number of round huts, built loosely of stone, and shown by excavation to have been occupied about the time the Romans were engaged in the conquest of North Wales. Within them charcoal was found, showing that fires had been lit, and iron and bronze implements, porcelain beads, and some bits of Roman pottery were also unearthed. The spot was, of course, too bare and bleak to have been permanently inhabited; it was, no doubt, a tribal retreat used in case of emergency in the summer. The great fort on the summit of Penmaen Mawr is of the same pattern; it has, also, well-built walls and many round stone huts, known to the Welsh as *cytiau*. The same holds good of Pen y Gaer, near Llanbedr y Cennin in the Conway Valley, and Caer Seion, on Conway Mountain, all belonging, there can be little doubt, to the tribes whom Agricola found in possession of Snowdonia. There are other forts, of which the history has not yet been unravelled, near Llanddeiniolen, on Carn Bentyrch and Carn Fadryn, at Porth Dinllaen, and in the neighbourhood of Llanwnda.

Both of the Roman forts situated in Carnarvonshire have yielded remains, as has the line of road which connected them. Caerhun has not been excavated to any great extent; indeed, the parish church and church-yard occupy much of the enclosure; but enough has come to light to show that it was a small military station, occupied by a detachment of the twentieth legion. The ruined building between the fort and the river was a bath-house, heated from below in the usual Roman fashion. Carnarvon was a somewhat larger fort, covering about $5\frac{1}{2}$ acres; it was placed on the hill which rises between Llanbeblig church and the modern town, and the vicarage stands within the ancient walls. As yet there has been no systematic digging over the whole site, but incidentally a good deal has been brought to the sur-face, illustrating the history of the place in Roman times. The most notable find was an inscription, now in the Castle, recording the fact that about 200 A.D. a cohort of auxiliary troops from the neighbourhood of the Meuse, the Sunici by name, repaired a broken aqueduct. No doubt, they formed at the time the garrison in charge of Segontium. Roman coins have been found, some belonging to the beginning and others to the end of the occupation of Britain, so as to suggest long possession of the district. On the road from Caerhun to Carnarvon four Roman milestones have at various times been discovered, two above Gorddinog, one at Ty Coch, Bangor, and one at Llanddeiniolen. The Gorddinog stones, now in the British Museum, marked the eighth mile from Conovium; they were of different dates and separated by an interval of about eighty years.

There are few remains of the long period which intervened between the withdrawal of the Romans and the Norman invasions. Building in stone went out of fashion and the arts were generally at a somewhat low ebb. Nearly all that the early Christian culture of Carnarvonshire has handed down as evidence of its existence is a number of Latin inscriptions, carved in stone as memorials of the dead, and preserved, for the most part, on sacred sites. There are, altogether, fifteen of these inscriptions, of which the most noteworthy are those of Penmachno, Cefn Amwlch, Llannor, Llanael-haearn, Cesail Gyfarch, and Llystyn Gwyn. The last-mentioned (discovered in 1902) has, in addition to the Latin capitals, an inscription in what is known as the Ogam character, which is common in Ireland and South Wales, but otherwise only occurs once in North Wales. All belong to the fifth, sixth, and seventh centuries and are definitely Christian in character; two (at Penmachno

and at Treflys) have the *Chi-Rho* monogram, i.e. the first
two letters of *Christus*, two commemorate priests and one
(at Llannor) is believed to record the name of the founder
of Llangwynodl church.

It is singular that Carnarvonshire, though otherwise a
happy hunting-ground for the archaeologist, contains no
specimen of the so-called "Celtic" ornamentation found

Treflys Church

in other parts of Wales on old ecclesiastical sites and
ascribed to the eighth, ninth, and tenth centuries. There
are good specimens at Penmon, just across the Straits.
On the other hand, the county possessed until recently
one of the very few Welsh specimens of the "Celtic"
quadrangular bell ; it belonged to Llangwynodl and is
now in the Welsh National Museum.

18.   Architecture—(a) Ecclesiastical.

The early churches of Wales were mostly built of wood, and it is not surprising that Carnarvonshire, in common with the rest of North Wales, should possess no architectural remains which can be referred to an earlier date than the twelfth century.   It has many quaint, old-world churches, harmonising well with their romantic surroundings, but few of these are really old, as churches go, and most of the work is of the fifteenth and sixteenth centuries.

Probably the oldest ecclesiastical edifice in the county is the north aisle of Aberdaron old church.   The church consists of two parallel aisles, separated by arches of the Perpendicular period, and is entered by a west door in the northern aisle, which is of the round-headed type, and thereby, as well as by the character of the ornament, is shown to be of the Norman period of architecture. There is another round arch on the north side of this aisle. Aberdaron is known to have been a very important church, possessing a good deal of land and maintaining a body of canons in the medieval period.   The building was probably erected in the twelfth century, when the place was famous as a sanctuary and a port for Ireland.

Another interesting old church is that of Conway. This has been since 1284 the parish church of the borough, but previous to that year it was part of the Cistercian abbey of Aberconwy.   The Cistercian order of monks, often known from the colour of their vest-

ments as White Monks, took their origin from the
monastery of Citeaux in France and in the twelfth
century spread widely throughout Europe. They prac-
tised a greater austerity than the monks of the old
Benedictine pattern, and discipline was kept up by regular
visitation of the various abbeys. The order became very
popular in Wales and the country suited their habits, for
they loved to settle in retired and solitary spots and had a
great liking for sheep-farming. Aberconwy was an off-
shoot of the famous abbey of Strata Florida, in Ceredigion;
it was founded in 1186 at Rhedynog Felen, near Carnarvon, but was moved a year or two later to the mouth of
the Conway. Llywelyn ab Iorwerth was its great bene-
factor and gave it lands in all parts of his dominions. A
good deal of the present building was originally built in
the monastic period, for instance the three lancet windows
of the western face of the tower, with the doorway below,
the south wall of the chancel, with its two two-light windows, and the two buttresses at the east end. All these
belong to the thirteenth century and are in the Early
English style of architecture. Other features of the
church belong to later periods, the east window and the
topmost section of the tower being of the fifteenth century.
This is also the date of the rood-screen, separating the
choir from the nave, a medieval feature of which there
are not many examples hereabouts.

The cathedral church of St Deiniol at Bangor occupies
a site which has been dedicated to ecclesiastical uses since
the sixth century, but the present building is not, as a
whole, of very great antiquity, and cannot compare in

interest or in dignity with the great cathedrals of England. There are some remains visible of the Norman church, in particular a blocked round arch in the south wall of the choir, but nearly all the older work belongs to the Early English restoration of Bishop Anian (1267–1305)

Bangor Cathedral

(with University College New Buildings in the distance)

or to the alterations made in the Decorated period (fourteenth century). The cathedral as it stood in 1402 was burnt to the ground by Owain Glyn Dŵr and for the best part of the following century it lay in ruins. Henry Dean, who became bishop in 1496, began the pious task of rebuilding; it is to him we owe the Perpendicular choir and

The Bangor Pontifical (illuminated page)

presbytery, with the east window. His work was worthily carried on by his successor Skeffington (1509–1533), who rebuilt the tower from the foundations and restored the nave and the transepts, thus giving the fabric its predominant character of a Perpendicular building. An inscription on the west face of the tower records how "Thomas Skevynton" caused this bell-tower and church to be built in the year of the Virgin Birth, 1532. The Reformation was followed by a good deal of neglect and, in the early nineteenth century, by injudicious alterations. Under Bishop Campbell (1859–1890), the whole building was carefully restored and except for the central tower completed, so that, if not an impressive cathedral, it is, at any rate, now in fitting order and repair. Its total length, measured within, is 214 feet.

Many famous men have been laid to rest in Bangor cathedral, such as Gruffydd ap Cynan, his sons Owain and Cadwaladr, and many bishops of the see. But there are no tombs of special interest, for that pointed out as Owain Gwynedd's is in the wrong position and can hardly be genuine, and the building has seen so many vicissitudes that it now contains few objects of historical value. An exception must be made in favour of the Pontifical, or service-book, of Bishop Anian, a manuscript of great interest preserved in the cathedral library. It was the bishop's manual, containing the liturgical forms he needed for the discharge of his various duties and recording, not merely the words, but also the musical notes of the anthems used in the different services.

At Clynnog Fawr there is a fine Perpendicular church,

The Bangor Pontifical (showing old music)

nearly 150 feet long.   Reference has already been made
to the antiquity of the religious foundation here, but there
is nothing to suggest any great age in the church, which
is entirely the work of the latter half of the fifteenth
century.   Among the interesting features of the building
are the rood-loft and screen, the former approached by
a spiral staircase in the thickness of the wall; the sedilia
or canopied stone seats for the clergy in the chancel ; the

Clynnog Church

upper rooms over the north porch and the sacristy; and,
most remarkable of all, the separate building, connected
by a passage with the church, which was known as Capel
Beuno or Cell y Bedd.   The detached building contained
the tomb of the founder, and at one time ailing children
were brought and placed upon it, to see whether they
would sleep, a sure presage of recovery.   Another institution of the place was Cyff Beuno, a very ancient wooden

chest, to which the farmers of Arfon paid tribute within the memory of men now living, to ensure the fertility of their flocks and herds.

There was an ancient monastic community of the Welsh pattern at Beddgelert, famous for its hospitable entertainment of the travellers in those wilds. At first, it belonged to no recognised order, but in later times it was treated as a priory of Augustinian canons, an order who, while taking upon them monastic vows, did not cut themselves off from the world, but performed clerical duties. During Edward I's conquest of Wales, there was a serious fire at Beddgelert and it was then, perhaps, that the church was built of which a good deal survives in the parish church of to-day. Of the early work one may note in particular the two beautiful arches on the north side of the nave, the severe but dignified three-light east window, and the western doorway.

Bardsey Island, the "Ynys Enlli" of the Welsh, was also a very ancient monastic refuge. Difficult of access by reason of the strong tides, it was greatly in request as a last retreat and resting-place for the medieval devotee, and Meilyr the Welsh poet expresses his fervent wish to be buried in "the fair isle of saints, in the midst of the heaving ocean." The air was reported to be so healthful that none died before his time, but each gave up the ghost as he became the oldest inhabitant. The life was, no doubt, simple and unpretentious, and all that remains of such buildings as the monks may have had is a ruined thirteenth-century tower.

The parish churches scattered up and down the county

have many points of interest, only some of which can be briefly alluded to here.   Llanengan (Perpendicular) has an extremely fine rood-loft and screen, such as would scarcely be looked for in so remote a situation.   Llangelynnin old church, remarkable for its situation on the bare mountain side, 900 feet above sea level, has a stoup

Llanengan Church

or holy-water vessel near the door, which is said to have been regularly used by the worshippers as late as the early part of the nineteenth century.   Dolwyddelan was built by Maredudd ab Ifan, ancestor of the Wynnes of Gwydir, who died in 1525; it has an old knocker, said to be a sanctuary ring (which fugitive offenders caught hold of), a rood-screen, and a brass tablet to the memory of the founder and his wife. Llangwynodl (Perpendicular) is remarkable as having three aisles, a rather unusual feature in North Wales.

19.　Architecture—(b) Military.

The northern half of Wales cannot compare with the southern in the number of its castles, for it was held by the Welsh, who were not great builders, until the age of Edward I, while most of the southern districts were ruled by feudal barons whose power rested on the possession of castles. Thus Carnarvonshire has only five medieval fortresses to show, viz. those of Carnarvon, Conway, Criccieth, Dolbadarn, and Dolwyddelan. Nevertheless, what is lacking in numbers is amply made up in distinction, for Carnarvon and Conway are among the most impressive ruins of the kind in the British Isles.

Upon completing his conquest of this district, Edward I chose Carnarvon, which had long been a residence of the princes of Gwynedd, as the seat of the chief castle and borough of North-west Wales. Work was commenced on the ground as early as 1283, but whatever was accomplished in the next ten years came to nothing in the revolt of 1294, when town and castle were destroyed. Operations were then resumed on a more substantial scale. The town was furnished with a ring of walls, a great part of which is still standing, and a new stronghold was erected, the building of which went on until 1322, in the reign of Edward II. The statue of Edward above the

Carnarvon Castle from the River Saint

main entrance was placed in position in 1320. The architect appears to have been a certain Master Henry de Elreton and the material used was limestone.

The castle is built on a rock which borders the river Saint and it is separated on the other side from the town by a deep ditch. Each of Edward's North Welsh castles was planned so as to suit the lie of the land and no two are identical in design. Carnarvon has been compared to an hourglass; it is long in proportion to its breadth, being about 300 feet by 120, and was originally divided into two portions by a wall across the middle. There are seven towers in the circuit of the castle walls, in addition to the gatehouse. Of these the most noteworthy is the beautiful Eagle Tower, overshadowing the mouth of the river, in which is pointed out the chamber in which, according to local tradition, Edward II was born in April, 1284. At that time, however, there can have been little of the castle above ground, and, although the prince was undoubtedly born at Carnarvon, this particular story must be dismissed as baseless. The Exchequer Tower, opposite the gatehouse, was the residence of the chamberlain of North Wales and official records were kept there. The Well Tower, between the Eagle Tower and the gate-house, contained the very necessary provision for fresh water. The Queen's Gate is a very imposing archway, which once gave access to the "Maes," or field, between the river and the town; it is the traditional scene of the presentation of the infant Edward as Prince of Wales to the assembled Welsh, but, as has already been pointed out, this incident must be regarded as pure legend. The castle has for many years been kept in excellent order and there has also been a good deal of conjectural restoration, so that internally it has not the air of romance of the ivy-grown ruins of Conway and Beaumaris. In 1911 it acquired a new association of interest as the scene of the investiture of Edward Prince of Wales, a ceremony never

Carnarvon Castle: Queen Eleanor's Gate
(*Scene of the Investiture of* 1911)

before carried out in Wales and planned on a scale of the utmost magnificence.

The second important position fortified by Edward I was the mouth of the Conway. Degannwy, on the right bank, was the ancient stronghold of the district, but it did not give control of Snowdonia, and accordingly

Edward crossed the river, ejected the monks of Aber-conwy, and planted his new borough and castle on the site of the Cistercian abbey. The building of Conway Castle went on at the same time as that of Carnarvon,

Conway Castle: Interior

but, as it did not suffer in 1294, it was completed earlier ; indeed, Edward spent in it the Christmas of that year. Like Carnarvon, Conway was provided with town walls, and these have been remarkably well preserved, forming, with the castle, a triangular or harp-like enclosure which is intact to this day. The gates were known as Porth Uchaf (Upper Gate), Porth Isaf (Lower Gate), Porth y Felin (Mill Gate), Porth yr Aden (Curtain Gate), and Porth Bach (Little Gate).

Conway Castle resembles that of Carnarvon in its irregular oblong shape and it is of about the same size. There is no gatehouse, for the main entrance was approached by a causeway from one side, leading to a barbican and a small platform in front of the gate. In the outer or western ward the chief feature is the great hall, 100 feet long; two only of the arches of the roof remain, but their ivy-draped ruins invest the scene with an air of singular charm. There was no well, and water was stored in a tank cut out of the solid rock. At the east end of the hall was the chapel, separated from it by a light partition only. In the inner or eastern ward were the state apartments, having a platform at the further end which overlooks from a great height the estuary of the Conway. Eight massive round towers, of identical pattern, surround the castle and preserve intact to this day the picturesque battlements with which they were crowned by their first builder.

Criccieth Castle is a much more modest structure than either of the two just described. It is known that the princes of Gwynedd had a stronghold here and there can be little doubt that Edward I merely adapted the ancient fabric to his own purposes. It occupies a striking situation, on the crest of a little hillock (the "crug" of the name) which is almost surrounded by the sea; two bold towers flank the entrance and provide the chief strength of the enclosure.

Dolwyddelan and Dolbadarn have been reduced to so fragmentary a state that it becomes difficult to tell their story with any degree of confidence. Neither plays an important part in history and all that can safely be said is

Dolbadarn Castle

that Dolbadarn seems to have been built about the time of Edward II to guard the plain of Arfon from attacks by way of the Pass of Llanberis, while Dolwyddelan, similarly defending the Conway Valley, may be in part a native Welsh building of the thirteenth century, but was much altered by Maredudd ab Ifan when he came to live there in the sixteenth. The square tower, standing on its rocky knoll, is the oldest part of Dolwyddelan and perhaps was part of the castle which Edward I is known to have found here in 1282.

20. Architecture—(c) Domestic.

The houses of Carnarvonshire, both old and new, are for the most part built of stone, of which there is an abundant supply, and roofed with slate, that other characteristic mineral product of the county. Scarcely any of the old half-timbered houses are to be seen which so pleasantly diversify the landscape in the well-wooded counties of the border; brick, also, is not in favour outside the towns. Even in the remotest districts, slate has displaced the old thatched roof, the older work being distinguished from the newer by the use of a smaller and rougher slate, certainly more picturesque than the smooth, blue slate of modern commerce.

Of the older houses of the county, none can compare as an elaborate memorial of a bygone age with Plas Mawr, Conway. It is a true town-house, hemmed in on all sides by the dwellings of plain citizens, yet built on a lordly and dignified scale. It was built, or possibly developed out of an older building, by Robert Wynne (1520–1598), an uncle of Sir John Wynne of Gwydir, and is a good specimen of a gentleman's house of the Elizabethan period. It consists of an entrance lodge, a lower court with a stone staircase, a middle block containing the banqueting hall, an upper court, and the north block (the oldest portion) with the

Plas Mawr, Conway

state apartments.   The interest of the building is greatly heightened by the excellent state of preservation of the interior; the oak panelling, the carved chimney-pieces, the decorative plaster ceilings, the heavy furniture, all help to carry the mind back to the era of its foundation in the sixteenth century. In "Queen Elizabeth's Room"

Plas Mawr: Queen Elizabeth's Room

the arms and initials of the great queen are displayed upon the chimney-piece, and throughout there is the greatest profusion of heraldic and other ornament, no two ceilings being alike in design. External features of interest are the stepped gables, the tower with its fine view, and the lantern window projecting in an angle so as to give light to the courtyard. Plas Mawr is the property of Lord Mostyn, but is leased by the Royal Cambrian Academy, who hold in it their annual exhibition of pictures.

The house of Gwydir, not far from Llanrwst, has famous associations, for it was the home of Sir John Wynne, but little remains of the original mansion and the place is now chiefly remarkable for the curious antiquarian relics it contains. Not far from Aberdaron, at the extremity of Lleyn, is the farmhouse of Bodwrda, which still preserves the graceful outlines of a well-built Jacobean country house. Cochwillan, where the Griffiths of Penrhyn lived before they moved to the better known house, can still show the hall built for the family in the fifteenth century, which to-day does duty as a barn. Vaynol Old Hall is a Tudor mansion, and near it is a domestic chapel of the same period.

There are a few old houses in Carnarvon and Conway of the civic type, such as the Vaynol Arms, Palace Street, in the former town, and the Aberconwy Temperance Hotel in the latter, but the Carnarvonshire towns have been for the most part rebuilt within the last 150 years and have little domestic architecture to show of any great antiquity. Many of the farmhouses, on the other hand, though simply and plainly built, with an eye to solid strength and protection from the weather rather than lightness and grace, are extremely old and illustrate bygone methods of construction. Such are Cymryd near Conway, Iscoed Isaf in the Nantlle valley, Bwlch Gydros near Llangelynnin, and Hafod Lwyfog in Nanhwynen, the last mentioned built in 1638.

21. Communications — Past and Pre= sent. Roads and Railways.

The earliest line of road laid across Carnarvonshire would seem to have been that from Conovium to Segontium, passing through Bwlch y Ddeufaen, which has already been more than once mentioned. It connected the Roman settlements of the east with the shores of St George's Channel and was the precursor of the later routes to Ireland. At a later period, probably, of the Roman occupation, a branch was thrown off to southward, which may still be traced from the neighbourhood of the Miners' Bridge on the Llugwy across the hills to Pont y Pant and Dolwyddelan, and thence up Cwm Penamnen to the county boundary. The objective of this road was, no doubt, the military station at Castell Tomen y Mur, near Trawsfynydd. Like many Roman roads in Wales, this was known as "Sarn Elen," i.e. Elen's Causeway, and a legend was told of its making at the behest of a mythical Queen Elen of the Hosts, who married the Emperor Maximus.

Little attention was given to road-making in medieval Wales and it was rather the Welshman's object to make access to his mountain refuges as difficult as possible. It was not until the country had been thoroughly subdued to English rule that regular ways of communication with the new settlements around the castles were set up. Even then, there was hardly any wheeled traffic ; land carriage was chiefly by means of pack-horses, driven along narrow mountain and forest tracks; and wherever a waterway was available for transport purposes, as for instance from Trefriw to Conway, advantage was taken of it. A good deal of importance attached to the ferries across the Menai Straits, of which there were five, at Aber Menai, Talyfoel (near Carnarvon), Moelydon, Garth, and Beaumaris. The first two belonged to the crown, in succession to the princes of Gwynedd, and were sold to the borough of Carnarvon in 1874. Moelydon and Beaumaris were also crown property, but the latter was acquired by the corporation in the reign of Elizabeth and formed a considerable element in the prosperity of the place. Garth ferry was an appurtenance of the bishopric of Bangor until purchased by the borough of Bangor at the time of the erection of the promenade pier.

A royal post to Ireland existed as early as 1574. It passed through Chester, Denbigh, Conway, and Beaumaris to Holyhead. Sir John Wynne thus described its course from Dwygyfylchi to Llanfairfechan in a tract written about 1625. "The way, beginning at the seashore within the parish of Dwygyfylchi, is cut through the side of a steep, hard rock, neither descending nor ascending until you come to Seiriol's Chapel, being about a quarter of a mile from Clipyn Seiriol, and all that way is two hundred yards above the sea, over which if either man or beast should fall, both sea and rock, rock and sea, would strive and contend whether of both should do him the greatest mischief. And from the chapel aforesaid the way is cut through the side of a gravelly, rocky hill, still descending until you come again to the seashore within the parish of Llanfair." The road, which in places was hardly three feet broad, was kept in repair by a hermit, who lived on the alms of wayfarers and an annual collection in the neighbouring churches.

From that day to this the road round the precipice of Penmaen Mawr has been a matter of anxious concern to those responsible for the safety of traffic along the coast of Carnarvonshire. After many attempts to improve the old road, Parliament voted £2000 in 1769 for the making of an entirely new one, and a few years later this was constructed by John Sylvester at a considerably lower level than the former roadway, but still about 300 feet above sea level. From Aber to Beaumaris the route lay across the Lavan sands, and this portion of the road also had its dangers from the ease with which at nightfall or in a foggy atmosphere travellers might be cut off by the inflowing tide. At such times it was customary to ring the bells of Aber church in order to give guidance as to the right direction to be taken.

With the closer connection between Great Britain and Ireland which culminated in the Act of Union in 1800, the need of better provision for the Irish mail-coach traffic became urgent, and a new route was adopted which avoided the ferry at Conway and the dangers of Penmaen Mawr and the Lavan sands. The enterprise and public spirit of the first Lord Penrhyn laid down a road of the first class along Nant Ffrancon and the Llugwy valley, so that the mails henceforth came by way of Llangollen, Corwen, Pentre Foelas, Bettws y Coed, Capel Curig, and Bangor. An inn and stables were built at Capel Curig,

Holyhead Road and Llyn Ogwen

which was to be the half-way stage between Pentre Foelas and Bangor, and the crossing to Anglesey was removed from Beaumaris to Porthaethwy or Bangor Ferry, where the road ended at "Jackson's Inn" (now the George Hotel).

This new route not only brought about the rapid disuse of the Aber passage, but it also diverted from

Menai Suspension Bridge

Chester and Conway important traffic which had hitherto gone that way. The result was a concerted effort to improve the old or northern road, and in 1826 this line of communication was put upon an entirely new footing by the opening of the Conway Suspension Bridge, superseding the ferry, and the making of the Penmaen Bach loop, as a substitute for the toilsome ascent of the Sychnant

Pass. Both routes were at the same time relieved of the troublesome crossing at Bangor Ferry by the building of the great Menai Suspension Bridge, also opened in this year 1826. The Suspension bridges were the work of the gifted Scotch engineer, Thomas Telford, who applied in them a principle then entirely novel, namely that of suspending the roadway from massive iron chains resting on stone piers and secured at each end in the solid rock. The following figures will illustrate the magnitude of the task involved in the construction of the Menai Bridge: length from pier to pier, 579 feet; height of road above water, 100 feet; total weight of ironwork, 2186 tons; total cost, £120,000. Suspension bridges have since become fairly common, but the two over the Menai and the Conway were pioneer structures of this type and the former still retains its interest as a triumph of engineering skill which in no way injures but rather enhances the beauty of its natural surroundings.

Besides the Irish coaching route, there was another old line of communication in the county, namely, the road from Dolgelly and Harlech to Carnarvon. According to the maps and road-books of the seventeenth and eighteenth centuries, it entered the county by way of Traeth Mawr and thence ran through Penmorfa, Dolbenmaen, Garn, Llanllyfni, and Llanwnda. A new road has been constructed along a considerable part of this route, but the old one may still be traced in the neighbourhood of Pantglas, running parallel to it along the slopes of Mynydd Craig Goch.

The early years of the nineteenth century witnessed

great activity in road-making in this district. It was at this time that the road from Carnarvon to Beddgelert, past Bettws Garmon and Rhyd Ddu, was constructed, as also that from Carnarvon to Capel Curig, through the Pass of Llanberis. These two were linked together in 1805 by the road through Nanhwynen (or Nant Gwynant). Expectations that Porth Dinllaen might supplant Holyhead as the port of embarkation for Dublin led Mr Madocks, the builder of the Traeth Mawr embankment, to make a good road from Beddgelert to that roadstead, by way of Tremadoc, Criccieth, and Four Crosses. The main road from Carnarvon to Pwllheli, through Clynnog and Llanaelhaearn, is of the same period.

About the year 1830 the main road system of the county was complete. It enjoyed a few years of unchallenged supremacy and then began to be threatened by the advent of the railroad, which has continued to develop for the last sixty years and to extend into most of the inhabited areas of the county. As in the case of the high roads, Carnarvonshire owed its first railway to its connection with the Irish traffic. The Chester and Holyhead Railway, begun in 1845 and completed in 1850, followed the older coach route and provided the county with railway facilities from Conway to the Menai Bridge. Side by side with the suspension bridges over the Conway and the Straits, Robert Stephenson, the railway company's engineer, constructed two Tubular Bridges, no less remarkable as achievements in iron construction than the structures of Telford, though planned on an altogether different principle and, it must be admitted, scarcely bearing comparison with their companions in beauty of outline. The tubular bridges are rigid and may be described as iron beams, hollowed to admit of the passage of the trains. They rest, not upon arches, but upon stone piers. The Britannia bridge has a central pier, resting upon the Britannia rock in the middle of the Straits. Like the Suspension Bridge, it is poised 100 feet

Britannia Tubular Bridge

above the level of the water and thus offers not the slightest impediment to navigation at any point; the channel is 1100 feet across and the roadway of the bridge has a length of 1841 feet. Both here and at Conway the tubes for the up and the down line are separate, so that each bridge is really double. The Britannia bridge was opened for traffic in 1850, the Conway bridge, which is much smaller, in 1848.

The Chester and Holyhead Railway became in time a part of the great organisation known as the London and North Western Railway, the first of the railways of the United Kingdom in respect of its annual income, though second to the Great Western in the extent of its lines. This powerful company, using the Chester and

Conway Tubular Bridge

Holyhead line as its base, has in Carnarvonshire, as elsewhere in North Wales, provided a system of branch lines as feeders which covers a considerable area of the county. From Llandudno Junction runs the branch to Llandudno, conveying an enormous seasonal traffic to that well-known watering-place. Along the Conway Valley another branch runs from the same junction to Llanrwst, Bettws y Coed, and (through a tunnel over two miles long) the slate-quarrying centre of Blaenau Festiniog. Branches also connect Bangor with Bethesda, and Carnarvon with Llanberis, while from the county town a line runs south through the pass of Bwlch Derwyn to Cardigan Bay, connecting at Afon Wen Junction with the Cambrian system. The company's divisional engineer for North Wales is stationed at Bangor and the place has some importance as a railway centre.

The south coast of Carnarvonshire, from Portmadoc to Pwllheli, is served by another company, the Cambrian, which extended its operations into this district from Mid Wales in 1866. The headquarters of this line are at Oswestry and the Pwllheli trains connect with the company's main line at Dovey Junction, near Machynlleth. It has been the dream of more than one generation of railway promoters to continue this line to Porth Dinllaen, but the omens are not particularly favourable to any scheme of the kind.

In addition to these ordinary-gauge lines, there are two small lines of a special type in the county. The North Wales Narrow-Gauge Railway is of the "toy railway" pattern and was opened in 1877 to exploit the mineral wealth of the Snowdon area; it runs, with one short branch, from Dinas Junction, near Carnarvon, to Rhyd Ddu. An extension to Beddgelert and Portmadoc has been undertaken, but not as yet completed. The Snowdon Mountain Railway is a rack and pinion railway, of the type common in Switzerland, and was opened in 1896 to provide facilities for the ascent of Snowdon. It starts near the Victoria Hotel, Llanberis, and after an upward course of five miles ends a little short of the summit.

22. Divisions — Ancient and Modern. Administration.

The most ancient division of Carnarvonshire is into cantrefs and commotes, a division which goes back to the remote past of the country, ages before it came under direct English rule. Neglecting Llysfaen and Maenan, which were parts of the commote of Rhos Uwch Dulas, we may tabulate the old areas of the county thus :—

1. [Part of] Cantref of Rhos—Commote of Creuddyn.		
2. Cantref of Arllechwedd	,,	Nant Conwy.
	,,	Arll. Uchaf.
	,,	Arll. Isaf.
3. Cantref of Arfon	,,	Arfon Uwch Gwyrfai.
	,,	Arfon Is Gwyrfai.
4. Cantref of Lleyn	,,	Dinllaen.
	,,	Cymydmaen.
	,,	Aflogion.
5. [Part of] Cantref of Dunoding	,,	Eifionydd.

The whole, except Creuddyn, was reckoned as belonging to Gwynedd Uwch (i.e. Above) Conwy.

Without entering into much detail, one may briefly indicate the location of these areas. Creuddyn was the district east of the Conway and bounded by the Afon

Ganol. Arllechwedd lay between the mouth of the
Conway and the Glyders; in the valley of the former
it was separated from Nant Conwy by the Dolgarrog
stream, while on the side of Arfon the dividing line was
the little river Cegin, which flows into Port Penrhyn.
Of its two divisions, the "Upper" lay around Aber, the
"Lower" around Caerhun. Arfon lay along the coast
from Bangor to the Eifl and extended inland as far as
Nanhwynen, thus including the whole of the Snowdon
group. The river Gwyrfai, running from Rhyd Ddu to
Aber Menai, divided it into two commotes. Lleyn was
the western peninsula; the summits of the Eifl parted it
from Arfon and the river Erch from Eifionydd. As to
its commotes, Dinllaen was around Nevin, Aflogion in
the neighbourhood of Pwllheli, Cymydmaen in the ex-
treme west. It only remains to add that Eifionydd,
which historically was very closely connected with the
opposite region of Ardudwy, stretched from the Erch to
the Glaslyn and met Arfon at the upland pass of Bwlch
Derwyn.

It has already been pointed out that the county was
formed in 1284 by the grouping together of these earlier
areas. The new division did not, however, supersede
the old, for the latter were taken over as they stood to
form the "hundreds" of the new system of administration.
Thus it comes about that the ten commotes specified
above are the ten hundreds of · Carnarvonshire. The
hundred, however, though long an important unit in
the government of the country, has now ceased to be
of any great consequence and its place as a sub-division

of the county has been taken by other areas of recent origin.

For the purposes of representation in the House of Commons, the county is divided into (i) the Carnarvon Boroughs, (ii) the Arvon, and (iii) the Eivion divisions. The Boroughs include Conway (with Degannwy), Bangor, Carnarvon, Pwllheli, Nevin, and Criccieth, the latter two being parliamentary, but not municipal boroughs. The rest of the county not included in the Boroughs, is divided between the Arvon (east) and Eivion (west) divisions, the line of demarcation running from the neighbourhood of Port Dinorwic to that of Beddgelert. The effect of this division is to include most, though not all, of the quarrying centres in Arvon, and to make Eivion predominantly agricultural.

In all matters which are common to the county as a whole, more particularly education and the maintenance of main roads, the governing authority is the County Council, consisting of 16 aldermen (or co-optative members) and 53 councillors (or elective members). Elections to the County Council are held every three years, in the month of March, and the county is divided into 53 areas, of which each elects one councillor. The county offices are at Carnarvon. The county police force is under the control of a committee of 30, appointed as to one half by the magistrates of the county and as to the other half by the county council.

For purposes of local government, the county is divided into municipal boroughs, urban districts, and rural districts. There are three ancient boroughs, dating from the period

of conquest, namely, Conway, Carnarvon, and Pwllheli,
and one of recent foundation, for the city of Bangor was
only incorporated in 1883. Each borough has its mayor,
elected annually, and is governed by a representative body
thus constituted :—

	Aldermen.	Councillors.	Total.
Carnarvon	6	18	24
Bangor	6	18	24
Conway	4	12	16
Pwllheli	4	12	16

The urban districts are governed by councils and are
chiefly areas in which the town element is one of recent
growth. They are Criccieth, Bettws y Coed, Bethesda,
Llandudno, Llanfairfechan, Portmadoc, and Penmaenmawr. The rest of the county is divided between the
following rural districts, the councils of which are chiefly
concerned with questions of public health—Geirionydd,
Glaslyn, Conway, Gwyrfai, Lleyn, and Ogwen.

Another organisation exists for the relief of distress and
the care of the destitute, namely the Poor Law Union.
There are four such unions in the county, each with its
own workhouse, at Bangor, Conway, Carnarvon, and
Pwllheli. Three of them extend into neighbouring
counties, while the Portmadoc portion of the county
belongs for Poor Law purposes to the Festiniog Union,
and the Bettws y Coed portion to the Llanrwst
Union.

Ecclesiastically, the whole of Carnarvonshire west of
the Conway, except for small parts of the parishes of

Llanrwst and Yspyty Ifan, is in the diocese of Bangor, while the portions lying to the east of that river, except the parish of Llandudno, are in the diocese of St Asaph. The county is divided between the two archdeaconries, the archdeacon of Bangor having under him the rural deaneries of Arfon and Arllechwedd, the archdeacon of Merioneth those of Lleyn and Eifionydd. It will be observed that these ecclesiastical areas preserve the names of the old cantrefs and commotes, though these have been superseded by other divisions for most purposes of civil administration. There are 65 ecclesiastical parishes or districts either wholly or partially within the county, but the number of civil parishes is 79, for many of the small parishes, still treated as separate areas for such matters as the levying of the poor rate, have been grouped together for ecclesiastical purposes. Each civil parish having a larger population than 300, and not possessing a council as a borough or as an urban district, has a parish council, with power to deal with such matters as lighting, water supply, and rights of way.

23. Roll of Honour.

" Each land breeds men of might," according to a well-known Welsh proverb[1], and Carnarvonshire is very far from being an exception to the rule. Its mountain fastnesses have been a nursery of warrior heroes, of silver-tongued poets, of eloquent divines, of men of distinction

[1] Ym mhob gwlad y megir glew.

in many forms of human service. It will suffice here to mention a few of the outstanding names, omitting those of living persons, whose career cannot as yet with any finality be reviewed.

Nothing is known of the Snowdonian leaders who opposed the advance of the Romans or of those who fought against Cunedda and his Brythons, so that the earliest known name among the warriors of the county is that of Maelgwn Gwynedd, who died about 550 A.D. Degannwy was his principal fortress and residence, and he was probably buried at Eglwys Rhos. His prowess, his liberality, his talent for command are fully attested by his contemporary Gildas, though the stern monk speaks with severe upbraiding of his many offences against Christian morals. Another early figure of this district is Maelgwn's son Rhun, from whom Caerhun is supposed to have taken its name, and who led, according to tradition, the host of the cantref of Arfon to the North to avenge an invasion from that quarter. Rhun was a man of great stature, like his father, and the *Mabinogion* describe him as having red-brown, curly hair. He was, that is to say, of the tall, blond type to which so many Welsh princes belonged and which is reckoned Celtic rather than Iberian.

Many of the "saints," or monastic founders of the sixth and seventh centuries, were connected with the county. Engan of Llanengan, and Seiriol of Penmon and Penmaen Mawr, if the pedigrees may be trusted, were cousins of Maelgwn Gwynedd, and the former was King of Lleyn. The brothers Tegai and Trillo came, we are

told, from Brittany to found the churches which bear
their names, and with them came their sister Llechid.
Tudclud, the founder of Penmachno, Gwynodl, and
Tudno were also brothers; their pedigree makes them
sons of the king whose lands were lost to him by the
inundation of Cantref y Gaeclod.

Of the princes of later times, it is difficult to mention
one who ruled over Gwynedd and did not come into the
closest relations with this region of Eryri. Lleyn, Aber
Menai, and Bron yr Erw witnessed exciting episodes in
the career of Gruffydd ap Cynan; Owain Gwynedd,
Cadwaladr, Llywelyn the Great and his son David lived
and were buried in the county. The last Llywelyn won
his early victories in the district and spent here the last
years of his life, when the arms of Edward I had
restricted him to the country west of the Conway.

The age of the princes was succeeded by that of the
great landowners, among the most notable of whom in
this county were Sir Gruffydd Llwyd of Dinorwig, the
insurgent of 1322; Ieuan ap Maredudd, whose houses of
Cefn y Fan and Cesail Gyfarch were burnt by Owain
Glyn Dŵr; Hywel ap Gruffydd of Bronyfoel, knighted
on the field of Poitiers and known as "Syr Hywel y
Fwyall" (Sir Howel of the Axe); Maredudd ab Ifan, the
builder of Dolwyddelan church, where he was buried in
1525; the Griffiths of Penrhyn, chamberlains of North
Wales for three generations; and their kinsmen, the
Williamses of Cochwillan. Maredudd ab Ifan was the
founder of the Gwydir family, represented at the beginning of the sixteenth century by his great grandson, Sir

John Wynne (1553–1626), one of James I's first batch
of baronets, and an active and enterprising figure in his
day.   Sir John combined an interest in antiquarian and
in practical matters; in the first capacity he wrote a

Sir John Wynne of Gwydir

valuable history of his family, which throws much light
on the social condition of Carnarvonshire in the fifteenth
and sixteenth centuries, while he was greatly concerned
as a man of affairs in the development of the mineral and
commercial resources of North Wales.   He busied him-

self about the reclamation of Traeth Mawr, the mining
of lead near Gwydir and of copper in Parys Mountain,
and the establishment of the woollen industry in the
Conway Valley. He was masterful and grasping, and in
his relations with Bishop Morgan of St Asaph, whose
rise he had aided, he showed a strong disposition to
browbeat one who had once been his dependant.

Bishop William Morgan may well be regarded as the
most distinguished son of the county, at any rate in the
ecclesiastical sphere. He was born of humble parentage
at Ty Mawr, Wibernant, in the parish of Penmachno,
about 1540. With the aid of the Gwydir family, he
received a good education, graduating M.A. and D.D.
at Cambridge. While vicar of Llanrhaeadr Mochnant,
he undertook the translation of the Bible into Welsh,
only the New Testament having been hitherto accessible
in the native tongue, and this in the clumsy version of
William Salesbury. He overcame, not only the inherent
difficulties of the task, but also the obstacles placed in his
path by jealous foes, and the book appeared in 1588, soon
after the defeat of the Armada. It has remained ever
since the standard Welsh version of the Scriptures, later
changes having been quite insignificant, and has been at
the same time a priceless boon to the religious life of the
nation and a very important influence in the development
of the Welsh language and literature. Morgan was re-
warded with the bishopric of Llandaff in 1595, and in
1601 was translated to St Asaph, where he died on
September 10th, 1604.

Among the many prelates who have filled the see

of Bangor one may single out for special mention Lewis Bayly, author of *The Practice of Piety*, which, as "Yr Ymarfer o Dduwioldeb," was widely read in Wales; Humphrey Humphreys, a lover of Welsh antiquities and the patron of Ellis Wynne and Edward Samuel; Benjamin Hoadly, whose latitudinarian writings, violently disliked by the bulk of the clergy, led to the "Bangorian controversy"; and Zachary Pearce, the editor of Longinus. The list of deans includes David Daron, the counsellor of Owain Glyn Dŵr; Richard Parry, editor of the Welsh Bible of 1620; and Henry Thomas Edwards, a vigorous Welsh writer and ecclesiastical leader.

The Cochwillan family produced a notable member of the English episcopate in John Williams, who became bishop of Lincoln and lord keeper in 1621, and archbishop of York in 1641. Williams was prominent in the troublous times which preceded the Civil War and endeavoured to play the part of a mediator between the Puritans and the Crown. But he had little success in his intervention in politics and, after defending Conway for the king, died in retirement at Gloddaith in 1650. His tomb may be seen in Llandegai church.

The roll of Carnarvonshire poets is a long and honourable one, as one might expect from a region of such rare natural beauty. In the early part of the fourteenth century, Gwilym the Black of Arfon lamented the ill fortune of his master, Gruffydd Llwyd. Later in the same century, Rhys the Red of Snowdon (Rhys Goch Eryri), who lived at Hafod Garegog, near Beddgelert, effectively used the new metre known as the "cywydd" for nervous descriptive verse. Dafydd Nanmor is shown by his name to have come from the same district, but his work was chiefly done in South Wales. In the sixteenth century the county produced one of the shining stars of Welsh poesy in William Llŷn (1534–1580), who, however, spent most of his years of maturity at Oswestry, and indited his elegies and odes of praise to the gentlemen of Powys rather than to those of Gwynedd. Nevertheless, he sang occasionally to the magnates of his native county, to the lords of Gwydir and Bryn Euryn and Bodwrda.

After the Elizabethan age came a period of poetic decline, which was especially marked in Carnarvonshire, until the general revival in Welsh literature which was one of the features of the Romantic movement at the close of the eighteenth century. The county then produced once more a capable antiquary and bard in David Thomas of Waen Fawr (1760–1822), best known as Dafydd Ddu Eryri. By profession a schoolmaster, Dafydd served bardism well by keeping it in touch with the past and enabling the new poetry firmly to root itself in the old. He was followed in quick succession by a series of bards destined to win for the county a unique place in the story of Welsh minstrelsy. Robert Williams of Bettws Fawr (1767–1850) wrote, under the name of Robert ap Gwilym Ddu, in the free and the "strict" metres, verse of enduring merit; his hymns, in particular, being among the best in the language. His pupil, David Owen of Gaerwen, like himself a farmer in the district of Eifionydd (1784–1841), won for himself lasting fame under the name of Dewi Wyn o Eifion, showing in his ode to "Almsgiving" (Elusengarwch) both descriptive and metrical ability of the highest order. Next came Evan Evans of Trefriw (1795–1855), schoolmaster and curate, who took the title of Ieuan Glan Geirionydd

Dewi Wyn o Eifion

and has rarely been excelled as a writer of smooth and melodious Welsh verse. His lyrics and hymns have won a secure place for themselves in Welsh literature. Ebenezer Thomas (Eben Fardd) was another school-master-poet (1802–1863), who spent his life in the quiet village of Clynnog and wrote noble odes, of which the best known is that on "The Destruction of Jerusalem." These are but the leading names in a galaxy of poetic talent which made the county about 1830 the indubitable headquarters of the Welsh Muse.   Of a later generation

Ieuan Glan Geirionydd

were William Ambrose of Portmadoc (Emrys), Independent minister and poet (1813–1873), whose masterpiece is the ode to "The Creation"; and William Williams of Carnarvon (Caledfryn) (1801–1869), who combined these functions with that of literary critic.

During the nineteenth century, Carnarvonshire produced many notable Nonconformist divines. John Jones¬ (1796–1857), born under the shadow of Dolwyddelan castle, settled at Talsarn as a Calvinistic Methodist preacher and attained a unique reputation for eloquence and pulpit power. The brothers Owen and John Thomas of Liverpool were brought up at Bangor, while Evan Herber Evans spent the best of his days at Carnarvon.

It remains to add a few names which fall outside the groups mentioned above. Such are John Owen, the epigrammatist (d. 1622), a master of polished Latin verse, who was born at Plas Du, near Chwilog; John Gibson the sculptor (1790–1866), born at Gyffin, near Conway; Griffith Davies of Cilgwyn, the actuary (1788–1855); and John William Thomas of Llandegai (1805–1840), superintendent of Greenwich Observatory. Gibson's name is one of the most distinguished in the history of British art. He was fortunate enough to attract the notice of William Roscoe of Liverpool, and thus the gardener's son became a pupil of Canova and Thorvaldsen and a sculptor of European renown. He spent the greater part of his life in Rome, but amid the many honours showered upon him had a warm affection for his native land. He was the last representative of the classical school of sculpture.

24. CHIEF TOWNS AND VILLAGES OF CARNARVONSHIRE.

Short Glossary of the Commoner Elements in Welsh Place Names.

Aber—the mouth of a river, whether on the coast or on another river.
Afon—river.
Bryn—hill.
Bwlch—pass, gap.
Caer—fort.
Capel—chapel.
Coed—wood.
Cwm—valley, combe.
Dol—meadow.
Eglwys—church.
Llan—monastery or church.
Llyn—lake (*pl.* llyniau).
Nant—valley.
Rhaeadr—waterfall.

(The figures in brackets after each name give the population in 1901, but in Bangor, Bethesda, Bettws y Coed, Carnarvon, Conway, Criccieth, Llandudno, Llanfairfechan, Penmaenmawr, and Pwllheli, the figures given are of the census of 1911.)

Aber (382), which is short for Aber Gwyngregyn, i.e. the Mouth of White Shell River, is a village at the mouth of a beautifully wooded glen, at the upper end of which is a graceful waterfall. Here the old coach route to Ireland formerly crossed the sands, which are extensive, to Beaumaris Ferry. In the middle ages the place was a residence of the princes of Gwynedd, and Princess Joan, wife of Llywelyn the Great, and her son, Prince David, died here. The castle mound may still be seen in the village.

Aber Falls

Aberdaron (1119) is the most westerly parish of the county. The coast opposite to Bardsey is wild and precipitous. In the village the only feature of interest is the Norman church, one of the oldest bits of architecture in North Wales, which stands on the very brink of the sea. The canons of Aberdaron were once an important body of clergy, who owned a good deal of the Lleyn peninsula. "Dick of Aberdaron" was a marvellous linguist of

Abersoch Harbour

the early part of the nineteenth century, who was familiar with all learned tongues, but was otherwise an uncouth eccentric.

Abersoch, a fishing village and summer resort, partly in the parish of Llangian and partly in that of Llanengan. The harbour is formed by the estuary of the river Soch. Not far off is Castellmarch, a very ancient homestead; associated with it is the legend of King March, who was said to have horse's ears, a

monstrosity which, like King Midas in a similar plight, he strove
in vain to conceal.

Afon Wen (White River), the junction of the London and
North-Western and the Cambrian Railways, commands a fine
view of Cardigan Bay. It stands on the ancient demesne of
Ffriwlyd, once the property of Conway Abbey.

Bangor (11,237) is the largest town in Carnarvonshire. It
owes its early importance to the episcopal see, founded in the
sixth century by one Daniel or Deiniol. The name means
"wattle fence" and, as other places in Wales and Ireland adopted
it, this came to be known as "Bangor Fawr yn Arfon," i.e. Great
Bangor in Arvon. When Wales was parcelled out into dioceses,
Bangor became the episcopal seat of Gwynedd, and it has ever
since been the ecclesiastical centre of the land west of the river
Conway. Nevertheless, it was but a village until at the beginning
of the nineteenth century the opening up of Lord Penrhyn's quarry
gave it importance as a slate-port. The diversion of the Irish
traffic from Beaumaris and the making of the Chester and Holyhead Railway followed, and the location here in 1884 of the new
University College for higher education in North Wales finally
gave the city a new position as the chief educational centre of the
northern half of the principality. In addition to the cathedral (for
which see p. 108), the public buildings include the new buildings
of University College, completed in 1911 at a cost of £100,000,
the Normal College for the training of elementary teachers,
founded in 1862 and taken over by the County Councils of
Carnarvonshire, Anglesey, Denbighshire, and Flintshire, the
St Mary's Training College, under Church of England management, and a block of government property which provides for
Post Office, County Court, and Inland Revenue business. The
science work of the University College is carried on in the old
College buildings, once a well-known coaching hostelry bearing
the name of the Penrhyn Arms Hotel. The bishop's residence is

now on the Anglesey side of the Straits and the old Palace has been converted into municipal buildings. The city is governed by a mayor, 6 aldermen, and 18 councillors; it is the youngest of the North Welsh boroughs, having been incorporated as lately as 1883. The Congregationalists and the Baptists maintain two colleges here for the training of their ministers, under separate management but with a common system of teaching. Of the outlying parts of the city, Glanadda is chiefly occupied by the

University College Old Buildings, Bangor

employés of the Railway Company, Garth and Hirael are marine in their interests, and Upper Bangor is residential.

Beddgelert (1230) lies at the foot of Snowdon, where the Glaslyn and the Colwyn meet in a luxuriant meadow, diversified by abundance of trees. The name signifies "Gelert's Grave" and has not as yet been satisfactorily explained, for the famous story of Llywelyn and his deerhound is a bit of folklore which has only been connected with this spot within the last hundred years and the grave was erected—sad to relate!—by a local innkeeper in order to give the legend a visible basis. Yet the place has genuine historical interest; it was the seat of a monastic community famous for its sanctity and for the virtue of hospitality, whose church still remains to testify by its size and proportions to the importance of the vanished priory.

Beddgelert Bridge

Bethesda (4716), an urban district in the parish of Llanllechid, is the home of the greater part of the workmen employed in the Penrhyn quarry. It is situated on the Holyhead Road, on the right bank of the river Ogwen and at the foot of Carnedd Dafydd. The village grew up around the Independent Chapel and hence derives its name. It has a secondary school under the Welsh Intermediate Education Act, and is the terminus of a branch railway line running from Bangor. Choral singing has always flourished in the district and in past years Bethesda choirs have won great distinction.

Bettws y Coed (925) owes its importance to the unique beauty of its situation and surroundings. Its name signifies "The Chapel of the Woods" and the suggestion of woodland seclusion is fully borne out by the place itself. Among the famous points of interest are the Swallow Falls (Rhaeadr y Wennol) and the Miners' Bridge on the Llugwy, the Conway Falls and the Fairy Glen on the Conway, and the Pandy Mill on the Machno. The Holyhead Road, coming down from the Denbighshire moors, here crosses the Conway on the Waterloo Bridge and then bends westward to Capel Curig. The parish has been formed into an urban district, which in respect of population is the smallest in the county.

Caerhun (987) is a large parish on the left bank of the Conway, of which the church stands within the walls of the Roman fort of Conovium. This was the point at which the Roman road from Chester to Carnarvon crossed the Conway; its westward course may still be traced over Bwlch y Ddeufaen.

Capel Curig, a little village at the junction of the Llanberis and the Bethesda roads to Bettws y Coed, is famous for its view of Snowdon, with the two lakes known as Llyniau Mymbyr in the foreground. The church is an ancient chapel of Llandegai; the hotel and stables were built about 1800 for the coaching traffic along Lord Penrhyn's new road.

Carnarvon (9119), the county town, is situated at the mouth of the river Saint, where it falls into the Straits. No place in the county is of greater historic interest. On the hill behind the town stood the Roman fort of Segontium, the terminus of the road from Chester; here was founded at an early period the church of Llanbeblig; and here the princes of Arfon had a royal residence down to the extinction of the native line of rulers.

Swallow Falls, Bettws y Coed

When Edward I conquered Gwynedd, he planted a castle on the rock which overhangs the Aber, placed beside it a walled town with borough privileges, and made the place the capital of North-west Wales. Since then Y Gaer yn Arfon (the Fort in Arfon) has always held a leading position in the district. It is a municipal borough, with a mayor, 6 aldermen, and 18 councillors. The harbour is managed by a Harbour Trust, whose authority extends over a considerable part of the Menai Straits. The Saturday market is a thriving institution, in which the farmers of Anglesey traffic with the quarry folk of the Carnarvonshire uplands. Among the public buildings are the county hall, the jail (which serves Anglesey and part of Merionethshire), the intermediate or county school, and the pavilion, originally erected for the National Eisteddfod of 1877 and since used on several occasions for this annual patriotic gathering, which elsewhere can be accommodated only in a temporary wooden structure. In the fine open space known as Castle Square is a statue of Sir Hugh Owen (1804–1881), pioneer in all the Welsh educational move-ments of the nineteenth century. The dominant feature of the town is the Edwardian Castle, one of the finest in Great Britain, now the property of the Crown.

Clynnog (1497) lies on the main road from Carnarvon to Pwllheli, but somewhat off the railway route. Its chief feature is St Beuno's church, a beautiful Perpendicular building which fitly expresses the importance of the place as an ecclesiastical centre from the earliest Christian ages. There are some remains of St Beuno's holy well and among the curiosities of the church is a pair of dog tongs or "gefail gŵn," for ejecting troublesome dogs from the sacred building.

Conway (5242) is correctly Aber Conwy, the mouth of the river Conway. A Cistercian abbey was founded here in or about 1186, of which the parish church of St Mary is a relic. Edward I dislodged the monks, providing for them a new home at Maenan,

Eisteddfodic Gorsedd in Carnarvon Castle

and built here a castle and a borough to hold the passage of the river. In later ages Conway Ferry became one of the links in the chain of communication with Ireland, and the Suspension and Tubular bridges emphasise to-day the importance of this connection. The special charm of Conway is the manner in which it has retained its old world features; not to speak of the magnificent castle, it has a complete circuit of walls, an Elizabethan mansion in Plas Mawr, and many other vestiges of antiquity.

Black Rocks, Criccieth

Criccieth (1376), a rising watering-place, stands on the shores of Cardigan Bay. The castle, crowning a rock which rises steeply from the sea, is in its present form the work of Edward I, but the princes of Eifionydd had a stronghold here in earlier ages. Criccieth was once a municipal borough, but it has long lost its privileges in this respect, although it is still included in the parliamentary boroughs.

L. C.   11

Degannwy is an outlier of the borough of Conway, being that portion which lies east of the river, in the parish of Eglwys Rhos. It has a pier for the Festiniog slate-trade, owned by the London and North Western Railway Company, and is gaining repute, especially among artists, as a summer resort. But the real fame of the spot rests upon the long history of the fortress which once stood upon the rock at the back of the village. The "arx Decantorum" was a Cymric stronghold in the days of Maelgwn Gwynedd; it was held by Robert of Rhuddlan, Earl Hugh I of Chester, John, Henry III, and the two Llywelyns, and only sank into obscurity when its place was taken by Conway.

Dolwyddelan (1112) is "The Meadow of Gwyddelan" (a name which also appears in Llanwyddelan, Montgomeryshire), and should not be regarded as if it were Dolydd Elen, i.e. Elen's Meadows. The castle, said to have been the birthplace of Llywelyn the Great, is a mile from the village and the sixteenth-century church.

Llanberis (3015) is a large parish in the heart of Snowdonia, well known for its two lakes, Llyn Padarn and Llyn Peris, and the romantic pass which leads from them to Capel Curig. The old church is in Nant Peris, between the pass and Llyn Peris, but the population of the parish is now for the most part concentrated at its north-west corner, where the ascent of Snowdon begins and where the quarrymen employed on the opposite side of Llyn Padarn have largely established themselves. Dolbadarn Castle is not far off. The Llanberis intermediate school is at Bryn yr Efail, in the adjoining parish of Llanddeiniolen. A branch of the London and North Western Railway connects Llanberis and Carnarvon.

Llanddeiniolen (6143) is a straggling parish, which includes a number of quarry villages, dependent upon the Dinorwig or Llanberis slate-quarry. Such are Bethel and Ebenezer (so called from Independent chapels), Dinorwig, Clwt y Bont, Bryn yr Efail, Pen Isa'r Waen, and Cefn y Waen. Dinas Dinorwig is a fine specimen of the British hill-fort.

Llandegai (2875) is the model village at the entrance to Penrhyn Park. In contravention of the usual Welsh rule, the accent falls, not on the penultimate, but on the final syllable of the name. The parish extends from the sea to Capel Curig and

Penrhyn Castle

Penygwryd, and includes Penrhyn Castle, Port, and quarry, the village of Tregarth, and the wild scenery at the head of Nant Ffrancon. In the church are monuments to Archbishop Williams and the first Lord Penrhyn.

Llandudno (10,469) is a name originally derived from the ancient church of St Tudno, which stands solitary in a fold of the tableland crowning the Great Orme's Head. At the beginning of the nineteenth century the name was applied to a small settlement of copper miners, who had squatted on the site of the present town. It was the opening of the Chester and Holyhead railway which directed attention to the possibilities of the spot as a bathing resort and its progress dates from 1850. The town had the advantage of being laid out from the first on a uniform plan, under the direction of the Mostyn estate. Its government was at first in the hands of Improvement Commissioners, but is now vested in an Urban District Council of 18 members. Llandudno is connected by a branch line with the London and North Western railway and an electric tramway maintains communication with the neighbouring watering-place of Colwyn Bay. Its many attractions for the pleasure-seeker are well known.

Llanfairfechan, or "Little St Mary's" (2973), at the foot of Penmaen Mawr mountain, is partly dependent upon the sett quarries and partly upon summer visitors. The old village, with the parish church, lies some distance from the shore and even from the Holyhead road; the growth of a small town in the intervening area dates from 1860, when the railway station was built.

Llanllyfni (5761) is one of the many rural parishes of the county containing a substantial urban element. The ancient church (St Rhedyw) on the banks of the Llyfni river is the centre of a small village, but the most populous place within the parish is Penygroes, which is the junction for the Nantlle branch. Other places within the parish are Nebo and Talsarn. The whole district is dependent upon the local slate-quarries.

Nantlle. The valley so called extends from Drws y Coed to Penygroes. The upper portion, except for a few mines, is rural and picturesque, forming a kind of antechamber to the Snowdon group; the lower is industrial and devoted to slate-quarrying. Two lakes once adorned the prospect, but the lower

has almost disappeared. Nantlle is Nant Lleu, the Valley of
Lleu, an early Celtic hero, who also gave his name to Din-lle(u)
and is identified with the Celtic divinity Lugos.

Nevin (1755) was the royal residence of the commote of
Dinllaen and was visited by Edward I, who held a tournament
here in 1284. The Black Prince made the place a chartered
borough, but it is now a borough only for parliamentary purposes.
In the past it has been a small shipping and shipbuilding centre,
but at present its chief importance is as a summer resort. The
beautiful coast scenery bids fair to make it very popular in this
respect. A motor omnibus service connects it with the Cambrian
railway terminus at Pwllheli.

Penmachno (1686) is a large parish at the head of the
Conway Valley. The true form of the name is Pennant Machno.
The church, dedicated to St Tudclud, is of very ancient foundation, as is shown by the number of early inscribed stones, dating
from primitive Christian times, which have been discovered here.
The father of Llywelyn the Great, Iorwerth Flatnose, was buried
in this sanctuary, and Bishop Morgan, the translator of the Bible
into Welsh, was a native of the parish.

Penmaenmawr (4042) is an urban district which represents the ancient parish of Dwygyfylchi. The parish lies between
the mountains of Penmaen Mawr and Penmaen Bach, but is
divided into two parallel glens or combes running seaward by
the intervening Moel Llus. In the eastern combe is the parish
church, with the hamlet of Capelulo; in the western the bulk of
the population, grouped around the railway station of Penmaenmawr. The place is of very recent growth and is in part a quarry
village and in part a summer resort.

Port Dinorwic is the village which has sprung up around
the harbour from which the slates of the Dinorwig quarries,
near Llanberis, are shipped. The Welsh name is Y Felin Heli

11—3

(The Saltwater Mill), from an old mill here which was turned by the ebbing tide. The ancient ferry of Moel y Don crosses the Menai Straits at this point. For purposes of local government, Port Dinorwic has no independent status, but is divided between the parishes of Pentir and Llanfair Is Gaer.

Portmadoc (4883) is the new harbour erected by Mr Madocks after his reclamation of the Traeth Mawr, which, as the result of the export of slates from the Festiniog area in Merionethshire, has become far more important than Tremadoc, intended at first to be the urban centre of this district. Officially, the urban district takes its name from the ancient parish of Ynys Cynhaearn; the parish church is three miles west of the town, Moel y Gêst intervening, and stands on an islet in the marshes of Llyn Ystumllyn. Portmadoc, in addition to its connection with the Cambrian railway, has communication with Blaenau Festiniog by means of a narrow-gauge line which was one of the earliest examples of its kind.

Pwllheli (3791) was the royal residence of the commote of Aflogion, being then known as Portheli. It was created a borough by the Black Prince, and is governed by a town council of four aldermen and 12 councillors. Pwllheli is the centre of an important agricultural district; it has a good harbour, which has of late been considerably improved. It is the terminus of the coast branch of the Cambrian railways. As happens not infrequently in the case of Welsh boroughs, the parish church is a little way out of the town; it bears the name of Deneio. The most important recent development in the history of the borough is its growth into some consequence as a watering-place. Of the two seaside suburbs, both of which front the waters of Cardigan Bay, South Beach is near the harbour and the Gimlet Rock, while West End is on the Llanbedrog side, being connected with that village by means of a horse tramway.

Trefriw (695), the " Hillside Hamlet," stands at the head of the navigable channel of the Conway, along which small steamboats ply to Degannwy. It is much resorted to in summer for the sake of the waters of the strong chalybeate well which is

Llyn Crafnant, near Trefriw

to be found a mile and a half to the north of the village. The houses are charmingly grouped on the steep face of a thickly wooded hill.

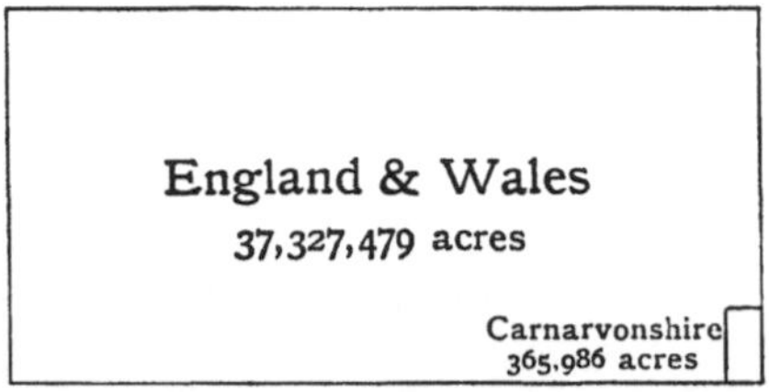

Fig. 1. Area of Carnarvonshire compared with that
of England and Wales

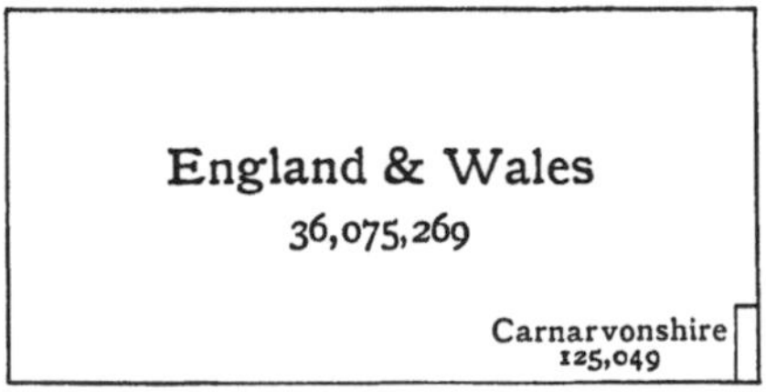

Fig. 2 Population of Carnarvonshire compared with that
of England and Wales in 1911

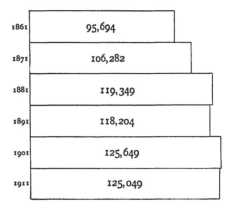

1861	95,694
1871	106,282
1881	119,349
1891	118,204
1901	125,649
1911	125,049

Fig. 3. Increase of Population in Carnarvonshire
from 1861 to 1911

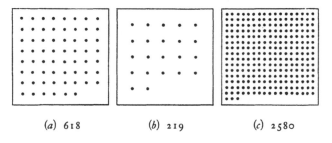

(a) 618   (b) 219   (c) 2580

Fig. 4. Comparative density of Population to the square mile
in (a) England and Wales, (b) Carnarvonshire, (c) Lancashire

(*Note, each dot represents 10 persons*)

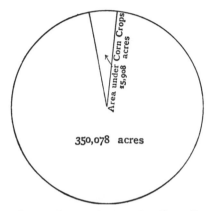

Fig. 5. Proportionate Area under Corn Crops in
Carnarvonshire in 1909

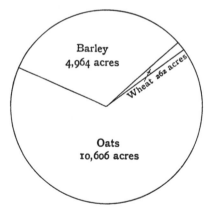

Fig. 6. Proportionate Area in Acres of Chief Cereals in
Carnarvonshire in 1909

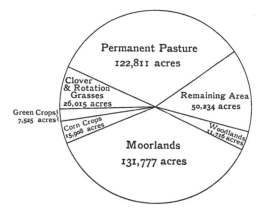

Fig. 7. Proportion of Cultivations in
Carnarvonshire in 1909

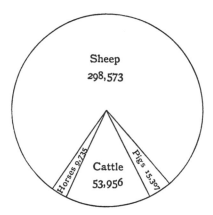

Fig. 8. Proportionate Numbers of Live-stock
in Carnarvonshire in 1909

Milton Keynes UK
Ingram Content Group UK Ltd.
UKHW041520181024
449640UK00009B/91

9 781107 641624